Career Options

Maria Sirdar-Nickel

Educator

By

Stanley R Taylor

Foreword by

Ray Bielecki
Assistant Director
Canadian Broadcasting Corporation.

Printed by Lighting Source Inc., Ingram Book Group, U.S.A.

Published by Stanley R Taylor

Published in Canada

Career Options: Maria Sirdar-Nickel, Educator:

Other books and articles by Stanley R Taylor

Taylor's Pneumatic Toys, LSI, Tennessee, December 2012

See www.stao.ca for articles

Maria,

You are an inspiration to children and adults alike. I feel privileged to have shared brief moments of your life as you move forward on your incredible journey.

Stan

This book is dedicated to all educators at home and in schools.

Table of Contents

Acknowledgements

Maria Sirdar-Nickel has been cooperative in sharing her numerous emails with me and in clarifying and adding more detail to different topics in the book.

Special thanks to Ray Bielecki for writing the Foreword.

Thank you to Maria's "buds" as she calls them and to the parent for their contributions.

A "Thank you" to my editor, Annette McLeod (www.mcleodwrites.com). Annette offered suggestions to the manuscript and I implemented most of them.

Thank you to Dorothea (writer@wsws.ca) and Rich Helms (rich@richhelms.ca) for their assistance.

Thank you to Paul Lima (www.paullima.com) for uploading my book to LSI.

Finally, thank you to my wife Karen and our family for their moral support.

Foreword

One particular lyric from *The Sound of Music* inspires me to write this Foreword: "How do you find a word that means Maria?"

Inspired, passionate, leader, driven, teacher, mentor, visionary, "cosmic-kid"…difficult to pigeon hole and perhaps that is a good thing because Maria is not a one-dimensional phenomenon. She is on a "mission" and has made lots of room for all kids on her cosmic educational adventure.

In the summer of 2009, my son Brett, who was seven years old, was quite the "space enthusiast". We had built "Brett's Space Ship-under-the -stairs", a 10 cubic foot area full of animated gizmos, space models, monitors and twinkling Christmas lights that were the window to our cosmos.

One night that summer, Maria was "teleported" into our lives as we were watching and an item came up by journalist Wab Kinew on CBC National News profiling the amazing space science education efforts by a teacher at Woodlands Elementary School in Woodlands, Manitoba….It was Maria!

We were captivated and inspired. I e-mailed Wab right away. My goal was to let Maria know about Brett's interest in "everything space" and I was looking for guidance on where I could receive educational space science resources because there was nothing available.

Within two weeks of my e-mail, in what seemed like "light-speed," a large manila envelope from Maria arrived. It contained a very inspiring and encouraging letter to my son "Commander Brett", a space craft colouring book, a book mark, a holographic card about the ISS, a pencil from the U.S. Space and Rocket Centre and the biggest treat was a packet of tomato seeds that were taken to the ISS by the Space Shuttle Endeavour. We found that "small gesture" or support from someone we had never met to be "out of this world"…but that's Maria.

Since that time, Maria has been a vital part of the AstroNuts Kids' Space Club, a group Brett and I founded to share our interest in space. Every year Maria inspires more and more elementary school children through her many activities and projects and our AstroNut kids always have lots of catching up to do during our Skype visits with her. Maria was especially engaging to over 350 very enthusiastic cosmic-kids at the 2014 3rd annual "What's Up in Space Camp & STEM Contest".

As a result of her passionate efforts, the Interlake School Division implemented the Students Spaceflight Experiments Program (SSEP), trained students and teachers how to compete in the program and to answer questions by Mrs. Nesbitt Fuerst and her Argyle students who came in first and who won an honoured place in the science labs of the ISS." Maria had also inspired an incredible team of kids at Woodlands Elementary School who placed second and third in the competition…but that's Maria.

 "How do you find a word that means Maria?"--Just look into the inspired and empowered eyes of the hundreds of smiling children that have met Maria and there lies your answer…. Thanks Maria!

Ray Bielecki
Assistant Director
Canadian Broadcasting Corporation

Introduction

When I taught elementary school, most of my students over the years had no idea what they wanted to do. I discovered that a large number of Grade 12 students were in the same dilemma. Students have been encouraged by educators to read about jobs held by others, but many of these were dated and to our youth irrelevant. Since I have interviewed 23 scientists/engineers over the past seven years, I decided to write a book about each one of them. This series "Career Options," will provide up to date information for our elementary and secondary school children hopefully providing them career choices they haven't considered.

***

Maria Sirdar-Nickel took my workshop on "How to Build a Pneumatically Controlled Canadarm" at the Space Exploration Educators Conference (SEEC) at Johnson Space Center (NASA) in Houston, Texas in 2013. This is how we met.

I took Maria's workshop on "Space Riot" at SEEC in 2014. We are friends and we follow each other on Twitter, Facebook and LinkedIn. I learned that Maria, as the Director of the Interlake Spaceflight Program, had to train 17 teachers and 450 students on how professional research was done. She also provided guidance on how to design an experiment like professional scientists, and to understand the environment of the International Space Station where the experiment was to be done. Such talent had to be shared and I decided to write a book about her. Maria is a multitalented woman as the following pages reveal. I am proud to call her my friend.

***

There are numerous Facebook quotes in the book. Those by Maria are simply followed by FB and the date. FB quotes by others are followed by the contributor's name, position, location and the date.

I have solicited comments from her friends and organizations with whom she worked. They are also followed by the contributor's name, position and location.

Most of the pictures in the book were taken by Maria's friends and relatives. Other pictures have been used with the permission of the organizations and individuals. I thank you all.

Some answers to questions I received from Maria came from an article I published in *Crucible* and *Elements* (the e-Journals of the Science Teachers Association of Ontario) entitled, "Careers in Science – Maria Nickel – A Passion for Science" Sept. 1, 2013.

Chapter 1 –Pre-School Years

Maria Anna Katharina Sirdar the daughter of Stan and Luba Sirdar was born two months premature on July 23, 1970. She had a 20% chance of making it. She was called the miracle baby due to the lack of prenatal innovations we have now. She was in an incubator for four months at the Royal Alexander Hospital in Edmonton, Alberta. She weighed in at 2 lbs. 15 oz. (1.34 kg) and was 21 inches (53.3 cm) long.

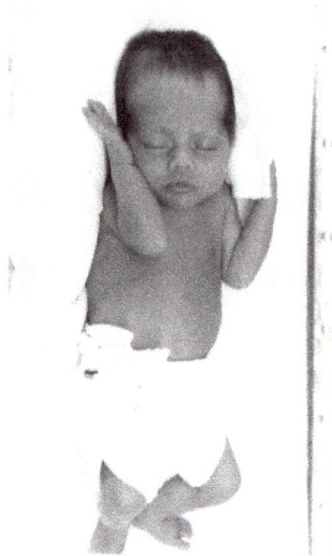

Maria in an incubator at two months old (left)

Maria at six months (right)

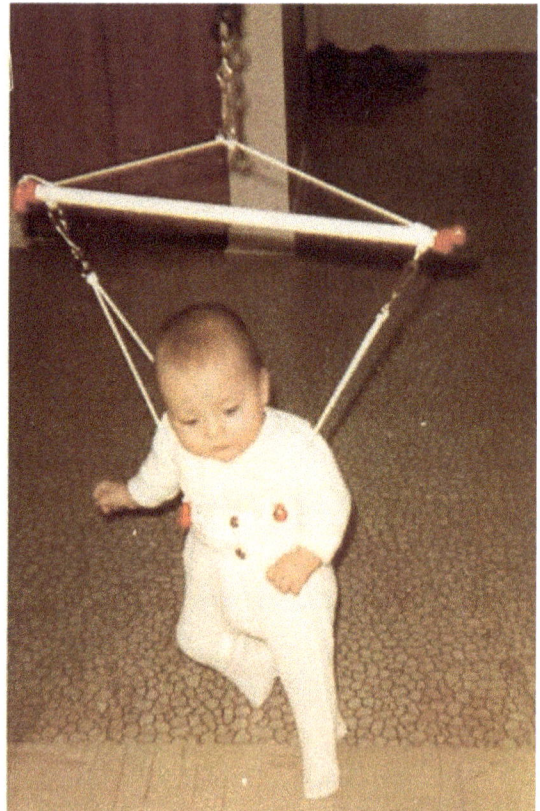

Maria is the oldest of six children having been predeceased by her older brother, Stanley Russell Jr. He would have been one year older than Maria. He was born at five months, weighed just 1 lb 5 oz, and he lived for five hours. His lungs were not developed enough to sustain him.

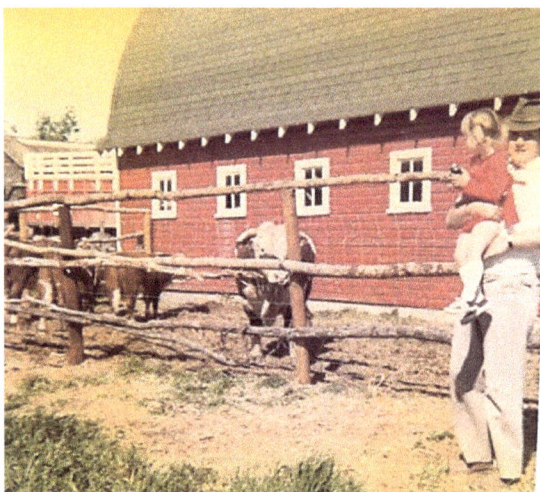

Maria age three on Dido's (grandfather's) farm with her dad (left).

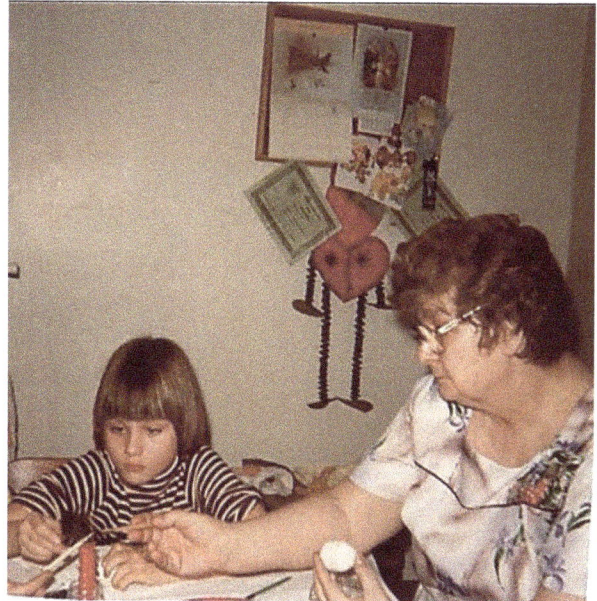

Grandma Sirdar and Maria painting Easter eggs (right).

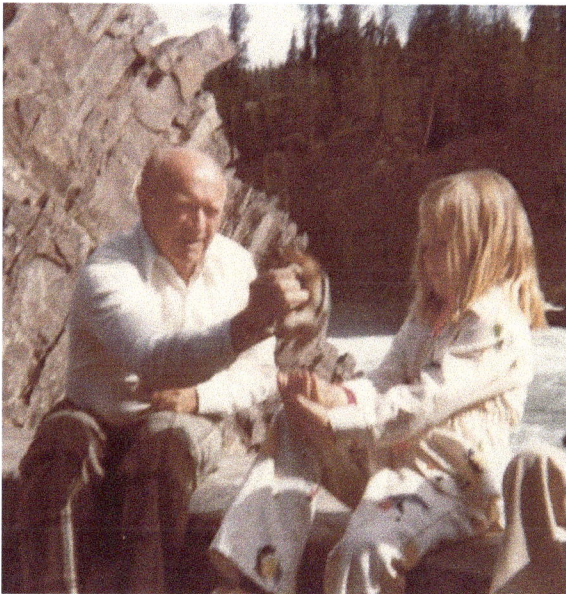

Bow Falls, AB with her Dido (left).

Maria at 3 ½ years of age (right)

She had a happy childhood playing with her friends, being at the park, swimming and spending time at the library. She loved hearing bedtime stories and she began reading as early as Kindergarten.

Maria age five

Halloween age six

Maria age seven

Chapter 2 - School Days

Kindergarten and Grade 1 were spent at Shamrock School in Southdale, MB which is now St. Vital School Division. Halfway through Grade 1, Maria transferred to Immaculate Heart of Mary Elementary, a private school which she attended through to the end of Grade 9. She went to St. Boniface Diocesan High School (private school) for Grades 10-12.

Modelling at age seven

Maria did Ukrainian Dancing from ages six to twelve

Her First Communion with her godmother, Auntie Marlene

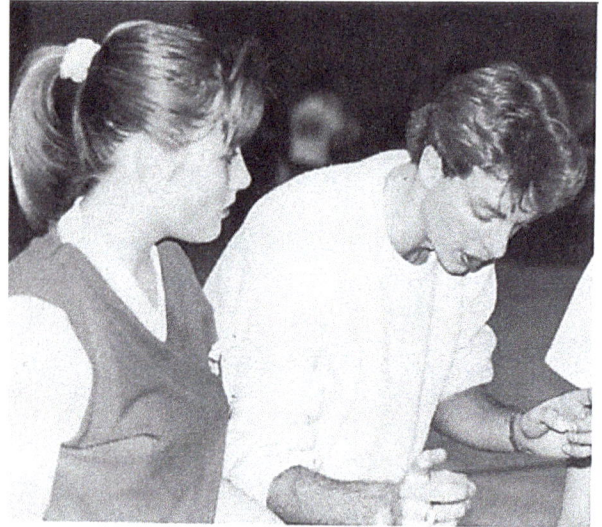

Listening to Coach Bob in varsity basketball.

Maria and Denis making pancake breakfast for their church youth group. Denis Radlinkski and Maria used to be Ukrainian dance partners and youth group members. She was his trainer for the University of Manitoba men's football team.

Maria, back row left, on the Social Justice Committee

5

Second from right, modelling in Fashion Show, Grade 11

Grade 12 Youth Parliament (back row left)

Varsity Volleyball (back row, second from right)

While in Grade 7, Maria was the Junior High Science Winner for a competition at St. John's School.

On the right is her individual plaque.

Gold Ribbons for her Science Project (left) on "How Oil Spills Affect Our Environment.

This win qualified Maria for the Provincial Science Fair called the Manitoba Schools Science Symposium (it still runs today). She got Bronze overall for her "Oil Spill" Project (picture below).

ENVIRONMENTAL SCIENCES ST. JOHN'S SCIENCE FAIR

PROJECT NUMBER: S5

GRADE: 9

PROJECT TITLE: HOW OIL SPILLS AFFECT OUR ENVIRONMENT

SCHOOL: IMMACULATE HEART OF MARY

NAME OF SPOKESPERSON: MARIA SIRDAR

NAME(s) OF OTHERS:

JUDGES INITIALS:

Maria became interested in science at a very young age. Her dad took her to see *Star Wars* and they watched *Star Trek* on television each week.

Female High School Athlete of the Year Award

Grade 12 graduation

Extra-curricular activities included swimming, volleyball, basketball, track and field, soccer, football, youth parliament club , badminton, year book club , social justice club, and modelling.

University of Manitoba Graduation

Maria Sirdar Nickel, Bachelor of Phys. Ed. with Athletic Therapy Major

Extra-Curricular Activities included swimming, beach volleyball, sponge hockey, slow pitch and orthodox baseball.

Maria graduated from Brandon University in Brandon, MB with a B.Ed. While she was there, she was the head trainer for the now defunct Brandon University men's hockey team the Brandon Bobcats. She dealt with all medical injuries, rehab, game prep, skate sharpening, equipment repair and all of the road travel items.

Chapter 3 – Space Academy for Educators

Space Academy, Huntsville, Alabama funded by Honeywell Corporation has its participants engage in activities in mathematics and science. The activities are in line with the U.S.A. Science, Math and Reading Standards. These activities are also of value internationally and they align with each country's science, math and reading standards.

"I AM SO IN, IN , IN , IN , IN! GOT MY LETTER OF ACCEPTANCE I AM GOING TO ADVANCED SPACE ACADEMY JUNE 18- 24TH, 2011, HUNTSVILLE ALABAMA AND FLORIDA HERE WE COME!!" (FB Mar 3, 2011).

Maria was one of 10 Canadians and one of 280 from around the world out of 1500 applicants to attend the Honeywell Space Academy in 2009. She was one of two Canadians out of 28

participants to attend the Honeywell Advanced Space Academy in 2011. At the one week Advanced Academy the training took place in two locations. The first part was at Huntsville, Alabama at the US Space and Rocket Center. The second part was at the Kennedy Space Center (KSC) in Florida. Maria states, "That was the year that Atlantis was the last shuttle to fly. My Advanced Academy

Team Opportunity

crewmates of 28 and I were on the platform launch pad at KSC. We did not see the actual launch. Atlantis was still being prepped. I watched the launch on TV with the rest of the world on July 8, 2011."

Maria at Cap-Com.

Maria in front of Atlantis.

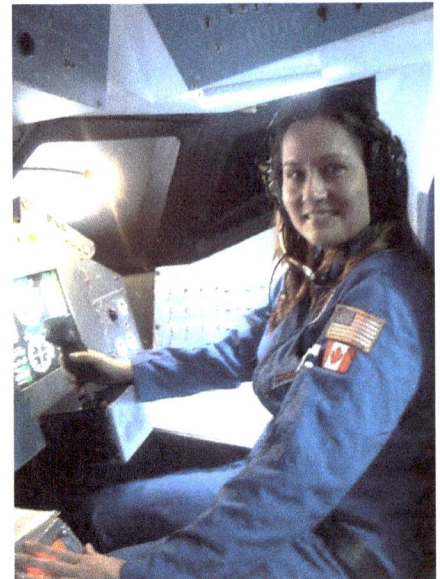

Maria in the Space Shuttle cockpit of Atlantis (right)

I asked Maria, "What was your first experience at Space Camp, Alabama like?" She responded, "Life altering! First PD [Professional Development] experience that changed the way I view science."

I asked, "What did you like most about the Advanced Space Camp?" She replied, "To have so many supportive, like-minded fellow teachers who encouraged my geekiness, love and excitement for space and to get such positive feedback and lesson ideas to share with my kids and colleagues back home in Canada was most advantageous."

In Zero-g plane with Kim Andrzejewski Astromovich at KSC

"Maria is amazing! We met at and attended Honeywell's Advanced Space Academy in 2011. We had one mission together where I was Commander, landing the orbiter, and Maria was the pilot, who supports the Commander. We were a phenomenal team! We worked very well together supporting each other to do our prep for the mission. I love the photos of us together in the cockpit of Atlantis! I was surprised to be chosen as Commander and Maria, with no negative ego, was so encouraging while I was so nervous to successfully land it. We kept each other calm and organized during this challenging time.

Maria was the initiator and designer of our team's Mission Patch. While others hesitated, she jumped in, became a great leader by organizing and getting everyone's input for the design. We all had something to add and say and Maria was the perfect person for the job.

Maria has the perfect blend of being serious when needed, but then she can be fun and goofy when you least expect it. She is hard working, devoted to her friends, students and family, and is an amazing role model.

Aside from our mission, I had a pretty bad allergic reaction to some meds I took for motion sickness (the bus ride from Huntsville to Kennedy Space center was very bouncy and made me sick) Maria used her EMT training to help me get through it. She consistently checked my neurological symptoms and heart rate to make sure I was safe. With some time I was better. I felt so secure with her calm, decisive and kind presence.

Since Space Academy we keep in consistent contact through texts, emails and Facebook." (Becky Loy, Special Education Teacher at East Syracuse Minoa Central Schools, New York).

"I met Maria (and Becky) when we were at Advanced Space Camp in 2011. She was as gung ho and amazing then as she is now. She was friendly to everyone and excited about everything we did. I think I was in mission control when she was in the shuttle, but she was always focused and was thinking about how to apply it all to her classes " (Jacqui Flowers, 5th Grade Math and Science Teacher, Arizona).

Space Academy Camp Entrance, 2009, Huntsville, AL

Saturn 1-B Rocket (below) and 1-B Stats (right).

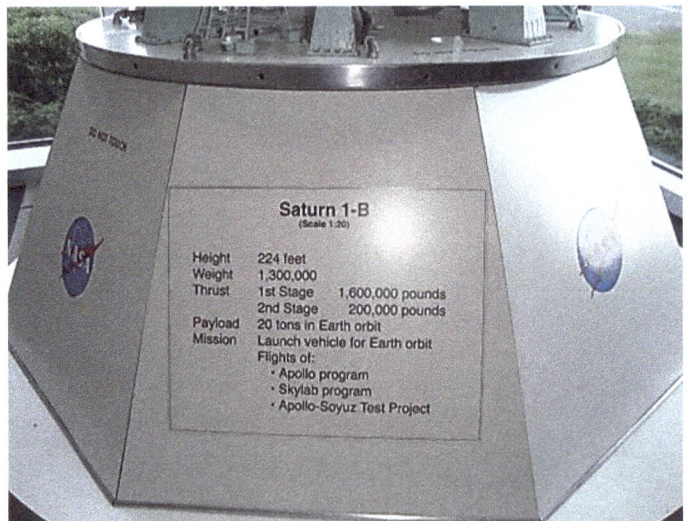

Saturn 1-B
(Scale 1:20)

Height   224 feet
Weight   1,300,000
Thrust   1st Stage   1,600,000 pounds
           2nd Stage   200,000 pounds
Payload   20 tons in Earth orbit
Mission   Launch vehicle for Earth orbit
           Flights of:
           • Apollo program
           • Skylab program
           • Apollo-Soyuz Test Project

Maria going for a spin (below and bottom right).

Advanced Space Academy group in front of the crawler platform that supports an Atlas 5 Rocket or the Space Shuttle

Maria as the ISS Commander, like Chris Hadfield was when he got to the ISS.

Maria (third from right) and her crew

Hanger for spacecraft of the Mercury, Gemini, Atlas and Apollo era (left).

Shuttle in the sling being hoisted to the crawler (right).

Shuttle, hydrogen tank, two rocket boosters on crawler ready to roll out to the launch pad. Atlantis took off on Friday July 8, 2011 at 11:28 pm (left).

The crawler (above and below) is the largest one in the world.

"This is the final draft of our mission patch from my drawing at camp. We are getting it made into a mission patch to put as a memory of our time as Advanced Space Academy team members" (FB July 12, 2011).

Chapter 4 – Space Exploration Educators Conference (SEEC)

Johnson Space Center (NASA) in Houston, Texas provides unique and challenging workshops and tours leaving recipients with information and lessons to enhance their classes. The opportunity to meet all the wonderful people from around the world, who share an excitement for space education, develops long-lasting friendships. This is what SEEC is about and it is worthwhile.

The Flight Suit picture taken by a photographer at JSC (left) makes a person look like an astronaut. Cool pic. Pricey, but cool.

"OMG so having a BLAST here at JSC Houston. Learned awesome activity for my kids on Osteoporosis, did Habitat NASA pods and I got to go inside and use the glove box and operate the mechanical lift and transport myself as the astronauts do when they train in this pod, to the second level in the living quarters they are testing to send to the moon or Mars to colonize , learned also how to build a vertical launching and gliding shuttle, a Geo Bat and solar system game and build the ISS model, and went to Kemau island for supper with buddies, and listened to a JAXA astronaut (Japan) on his adventures in space as a medical doctor, awesome and more to do tomorrow, so so so fun."Back from Houston, TX (2012). HAD A BLAST." (FB Feb 6, 2013).

2013 Gangnam Space Gals L to R Gangham prodigy Eric Sim the Intern, Becky Loy, Jacqui Flowers, Maria Nickel and former astronaut Clay Anderson, at NASA JSC. (Photo courtesy of Angela Case, Education Coordinator, Space Center, Houston). See http://goo.gl/93OKrP for "NASA Johnson Style" performed by Eric the Intern. For lyrics, see goo.gl/a4F6cI for NASA John Style at SEEC Dinner and Dance 2013 see goo.gl/l8G1xW

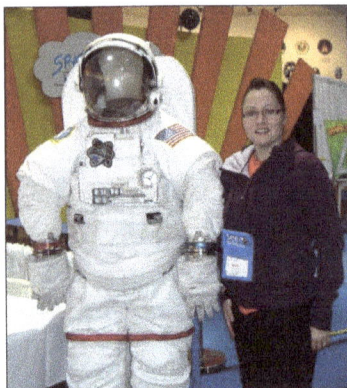

The STS 134 Spacesuit that will be used for Spacewalks in upcoming missions.

"Being this close to Gene Krantz (NASA Flight Director and remembered for his "Failure is not an option" quote in Apollo 13) was a privilege."

Brian Ewenson with Maria. Brian is an aerospace educator and professional speaker on the space program and aviation. He trained Chris Hadfield on his first mission to the ISS. Brian was born in Montreal, Canada and now resides in Texas.

Dive session with roomie Jacqui Flowers.

At SEEC 2014 on the Friday evening, there was a buffet dinner followed by a dance. The dance was a costume affair where those in attendance were asked to wear masks. Maria is flanked by Becky and Jacqui, her buds (left) with the amazing designs and the creative talent they put into their masks.

Chocolate Space Shuttle dessert following the meal (right).

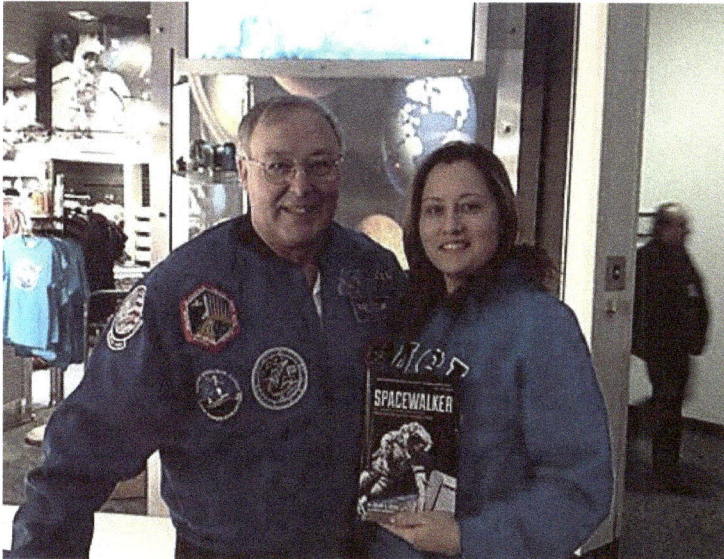

Maria met Jerry Ross, USA Astronaut and in May 2014, he was inducted into the Astronaut Hall of Fame. Maria bought Jerry's book, *Spacewalker* (a journal of his seven flights to and missions aboard the ISS).

Pasta Exploration Vehicles Session Winners with Jacqui Flowers (right).

While at home, Maria wrote on FB on Feb. 2, 2014, "A LOT EXHAUSTED at the moment! along with other stuff...and a lot to do: tests to mark, planning for sub for Wed while away at SEEC presenting & learning, software to download and teach teachers on Monday at school, problem solving lesson to finish up, unpack wash and repack for Wed flight to Houston, make sure all my presentation materials are in order, goodie bags for attendees

Maria doing her Space Riot workshop at SEEC 2014 with me at the back table flanked by two ladies.

and IN MY CARRY ON so that it does not get lost and & don't have it for Thurs presentation at SEEC . I think I would have a stroke if it were in my packed luggage & did not make it to Houston. /o/ I can live without clothes for a couple days (just buy some out there) but I can't be without my presentation materials."

Maria states: "It was a true pleasure to be in this picture with some incredible educators!" From left to right: Jennifer Cheesman, Becky Loy, Michael Wilkinson, Jacqui Flowers, Maria Nickel, Diana LeSueur at Space Center, Houston.

Upon returning home from SEEC, Maria wrote on FB on Feb. 11, 2014, "OK just saying!! I have the most wonderful amazing hubby EVER!! I came home from SEEC at NASA JSC & he surprised me with having the living room he has been working so hard on reinsulating, leveling, repairing, wiring & lighting up these last few months, by having it carpeted , painted up and had the new couch I bought in when I walked in . Still have to do baseboards and put pics up, but we can be in there finally after months. and TODAY he was off work, what does he do (I had late mtgs tonite till 6pm) he makes the last of the playdough I need for my space riot & 2 batches of cupcakes I needed to finish up for the riot for Friday when I got home. I LOVE HIM!! Sorry to blab , but it really made my day after the last 2 wks of crazy traveling, prepping for it , catching up from it, and a million other things on the go for work. It meant so much to me love ya Bobby !! u are the world to me!! — feeling wonderful."

Chapter 5 - The Space Club

Maria started the Woodlands Space Knights Club at her school in 2009 to help her kids learn about space. The activities she taught her kids she learned at the Canadian Space Agency conference in Longueil, Quebec and Space Academy and Advanced Space Academy in Huntsville, Alabama.

Maria writes: "Chris Hadfield did a live Skype from Kazakhstan, Russia with our entire school in the gym on Dec. 15, 2010, while he was in the middle of Training for Command of the ISS. The Canadian Space Agency arranged it. My space club kids got to ask a majority of the questions as well as a few Grade 8 students. I am sitting with Travis. You can see Chris Hadfield on the computer monitor. It was such a thrill."

Maria's Space Knights Club

The students did a number of experiments in the club.

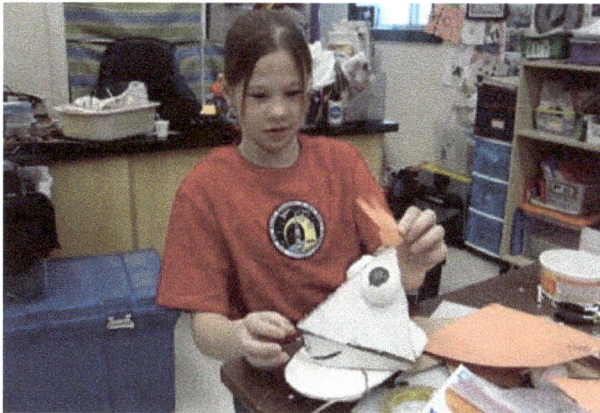

Laura (left) is doing the robotics unit. Students learned how to build their own remote controlled robot and the shell to cover it. Laura was doing Phineas from the TV show Phineas & Ferb.

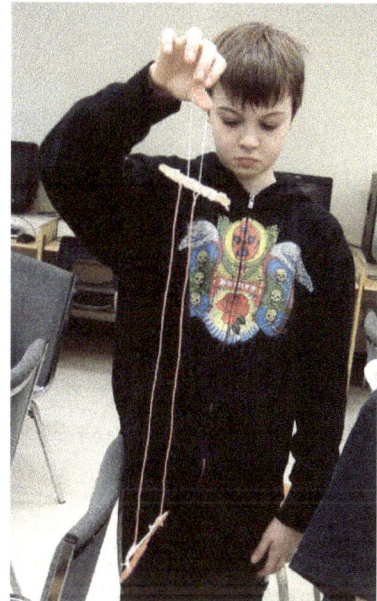

Evan Brad (right) was doing the toys in space lab on how gravity affects them. This experiment was done by NASA astronaut Leland Melvin. The students watched a DVD of these toys performing on the ISS in microgravity.

Space Knights robots making session.

The Space Knights Club taught the entire school how to build these toys as a Christmas activity. The whole school loved it.

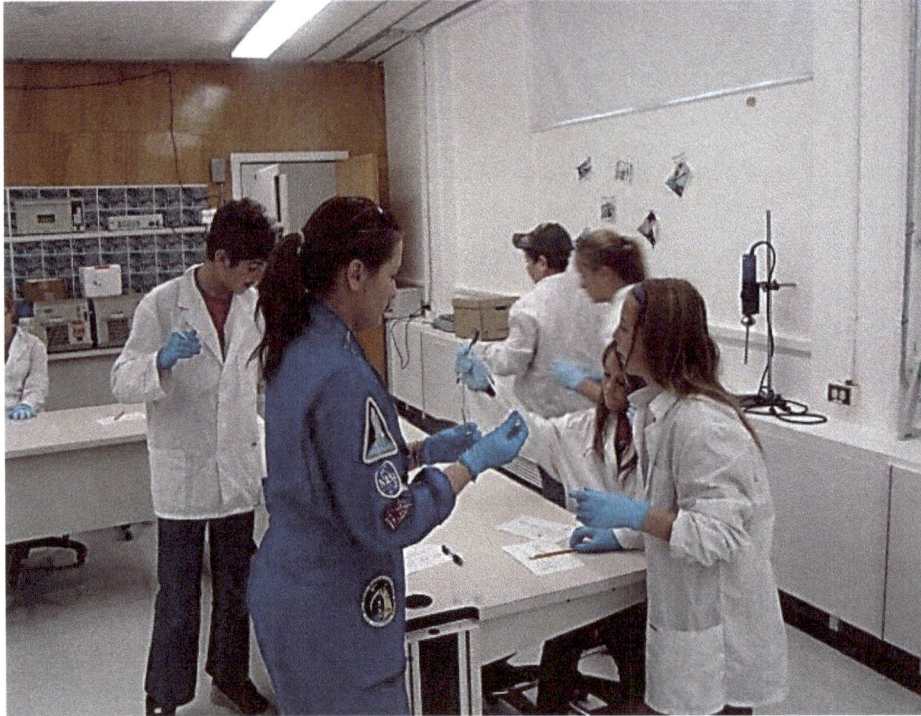

Maria's Space Knights Club went to the Inner City Science Lab during Biotechnology Week in Manitoba.

This new inner city science school with a state-of-the-art facility and equipment had their Faculty of Medicine grad students host Maria and her Space Knights with tours and all kinds of hands on experiments. In the photo above, the students were mixing chemical compounds and learning how to identify a mystery compound, a CSI type of investigation.

"The NASA curriculum I was working on with my test group (Space Knights Space Club) is now posted on their site. Go to the introductory page to see my name along with Jacqui Flowers , Becky Loy, and Amy S. Bartlett. Really cool experience. See goo.gl/L8di0W goo.gl/ETZM21 (pg 8 of document is LINK TO WORKBOOK). (FB Feb 15, 2014).

Chapter 6 - The Experiment

 "Well we have three finalists that are moving on to the last stage which is the big international panel in Capitol Heights Maryland USA!! I have one school called Brant Argyle with Mrs. Nesbitt Fuerst and a group of Gr 5/6 boys and girls who made it, then a group of Gr 5 girls from my school and Mr. Enns classroom, and the last one is from my room and a group of Gr 6 girls. SO thrilled. Today went so well! with two news stories in one broadcast. I could only find the one on line, trying to find the other one. Here is the link it is at 20:48 http://goo.gl/FggjMv (FB Nov 15, 2012).

"3 finalists being sent out tonite for last stage review panel in US. Yikes we are closer to knowing who will be the ONE!! So excited for the kids" (FB Nov 16, 2012).

Here is the link from the story on In our Canada on the Spaceflight Experiment Program. http://goo.gl/kzjB03 It is at the end 40 mins in, they did such a super job (FB Nov 21, 2012).

 "2 press releases done and ready to go out on 12th, patches in and ready for Houston, write ups in for patches for Houston, last bit for flight safety review being done, now to get sponsors for funding to send our crew to the launch hmmm" (FB Dec 6, 2012).

"Any teachers in Manitoba or Canada interested in going on this journey with your students to the Space Station and have an experiment, go to Student Spaceflight Experiments Program (SSEP): http://ssep.ncesse.org/ and contact Dr. Jeff Goldstein jeffgoldstein@ncesse.org He is accepting applicants for Mission #4!!!! Go for it People! Maria "(FB Dec 10, 2012).

"Today Interlake School Division was featured on front page of SSEP : http://goo.gl/3LW9wy (FB Dec 10, 2012).

"It is official, Interlake School Division chosen experiment and two mission patches have officially passed NASA flight safety review. We are the first international elementary school and Canadian one to do so!! SO PUMPED WE ARE GO FOR LAUNCH SPRING 2013!! c u Chris Hadfield in spring 2013 on ISS!!" (FB Dec 17, 2012).

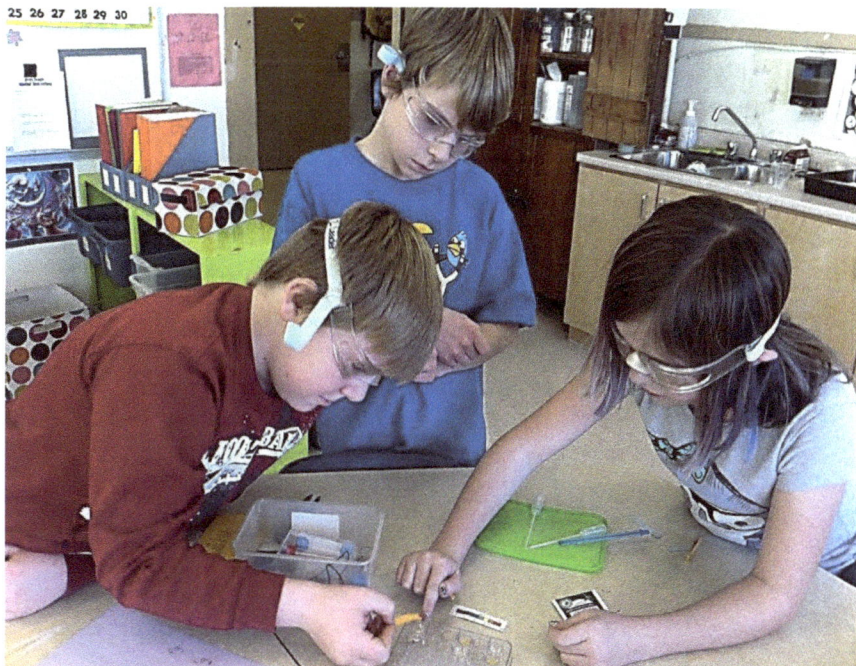

***

I sent an email to Leslie Nesbitt Fuerst, the Argyle students' teacher with the following request: "I want your input on your students' experiment. Who thought about the experiment? How much time did the kids put into the experiment including research and poster (I need hours, approximately). I know they did 95% of the work. I would like your views on their being selected to have their experiment go to the ISS."

Mrs. Nesbitt Fuerst replied: "The day I introduced the SSEP program to my Grade 4/5 class happened to be Terry Fox Day at Brant Argyle School. We had spent a great deal of time talking about what Terry Fox did to raise money for cancer research, what cancer is, what types of cancers there are, and how cancer has touched the lives around us. I began explaining the SSEP experiment to them and that they needed to design an experiment that would fit into a test tube. We wanted to see what effect microgravity at the International Space Station would have on the contents. Each team in my class had to come up with an experiment. The team of Ryan Petricig, Avery Good, and Ethan Enns wanted to know what would happen to cancer cells if they were sent to space. They thought that maybe the cancer would slow down because of microgravity. They thought that maybe microgravity could be a "cure." I had explained to the class that we were able to ask "partner researchers" or professionals for input on the experiment. I told the group that I would contact CancerCare Manitoba to see if we could obtain some cancer cells if our idea happened to be chosen. What I thought was quite a long shot unfolded rather quickly once I was put in touch with Eilean McKenzie-Matwiy, a Research Officer from the Institute of Cell Biology at CancerCare Manitoba. She recommended cervical cancer cells and it seemed that

our experiment was a go until we started to look more into the duration the experiment would be in space for, as well as the conditions that the experiment would be housed in. In the end, the first experiment that we came up with was not viable. We thought that our experiment was null and void until one of the partner researchers came up with the idea that we could send yeast to space. Without going into any further detail about the

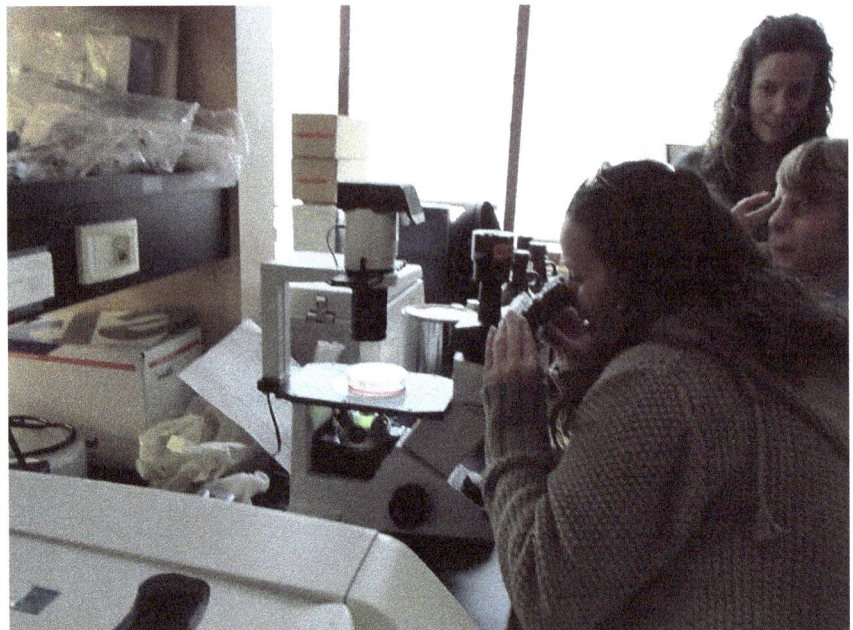

Leslie looking through the microscope at CancerCare with Ethan and Dr. Elizabeth Henson assisting.

29

experiment, that is how the idea was born. Then there was an addition of an anti-oxidant, which was originally going to be from broccoli, and that is how the experiment started to take shape.

We began approximately Sept. 9, 2012. The class had about two months to work in their groups. I would meet with them and tell them the revisions that would have to be made, and there were a lot: These are grade 4/5 students! Once their final experiment was submitted to me I had our principal select the best proposal to be submitted from our school. We then fine-tuned that one during their breaks before we submitted it to Maria. Once we found out that we won the competition, we had to put it into the format that was acceptable by the SSEP. This was hours upon hours of work during the kids lunch breaks, before they started school, and I would often pull them out of other classes to work on the project to get it finished by SSEP deadlines. The kids had to do research well beyond their years on cells, microgravity, cosmic radiation, etc. Most of the sources were not written in kid friendly language so I would try my best to interpret for them in a way that they could understand. In the beginning of November, we submitted a draft to SSEP, which went through a number of revisions. The final draft was submitted mid-December with minutes to spare. There were a lot of time crunches! It is hard to put a number on the actual hours we spent on the project and all that it entailed. Writing and researching alone were for sure over one hundred hours, along with discussing and emailing with the researchers, doing media events and explaining the project at various functions. The kids even made a movie trailer. (We were trying to interest Rick Mercer to no avail).

Just doing the project, before even being chosen as the experiment to go to the ISS, was a fabulous experience for the kids. They had the opportunity to do something real, that showed them what real science is about. I was ecstatic that my team of students was chosen. It was the first of many moments that made my heart swell with pride that my students could accomplish something so great. They are fantastic kids that I have known for many years. Watching them grow throughout the project has been amazing; seeing their career dreams change and be molded by this experience is something that every teacher aspires for."

***

"Congratulations to Arygle Gr 5 students and their teacher [Leslie Nesbitt Fuerst] for being chosen to fly their experiment to the Space Station in Spring 2013. SO proud to be a teacher and be involved in this project of SSEP. Minister Bjornson came out as well as some funders, ISD board members and only local newspapers, Not one came from any TV to cover the excitement at being first in MB & CANADA to do this, not impressed!! But the story will go out locally, so that is something at least, I was disappointed for the kids!! Nice thing was some of our sponsors said for me to call them to help out with sending the kids to the launch. I was not expecting that, yay" (FB Dec 17, 2012).

Bryan Petricig, Avery Good, Ethan Enns and Leslie Nesbitt Fuerst.

The honorable mentions for the Student Spaceflight Experiment Program went to Woodlands Elementary Gr 5 girls from Mr. Harold Enns experiment on "Algae Growth in Microgravity as a Biofuel Source" and four Grade 6 girls in Maria's class (these girls were also members of her Space Knights Space Club at school) whose experiment was on "The Effects of Microgravity and Cosmic Radiation on Honeybee Raw Royal Jelly and Can It Be Used as a Nutritional Supplement to Limit Bone Loss in Astronauts or Osteoporosis Patients."

Ryan Petricig pours a tea mixture from a flask into a beaker while Ethan Enns and Avery Good look on. Photo credit: Manitoba Institute of Cell Biology.

Cancer Care's Dr. Elizabeth Henson, Brant-Argyle students Bryan Petricig, Avery Good and Ethan Enns, experiment director Maria Nickel, Brant-Argyle science teacher Leslie Nesbitt Fuerst, and Cancer Care's Dr. Eilean McKenzie-Matwiy. Maria is dressed in her astronaut overalls having recently participated in the Advance Space Academy for Educators in Huntsville, Alabama.

Maria learned about experiments Chris Hadfield was doing through the CSA website and on his Twitter feed. She also learned about a contest put out by the Canadian Space Agency all across Canada back in October, 2012, where any school group of kids or individuals could participate in designing an experiment with the hope of having it selected to try out while onboard the International Space Station (ISS). She had to come up with an experiment that Chris could do with a list of materials already on the ISS.

 **** NOTE : this CSA contest was totally separate from the SSEP one. It was a total fluke that one was done at the same time we were doing the SSEP****[Maria Sirdar Nickel].

Maria stipulated at the beginning that the three students from the Brant Argyle School were from a different school than her own. Maria taught at the Woodlands School. Both schools are in the Interlake Division. Maria has also stipulated from the outset that the Brant Argyle students' science teacher Leslie Nesbitt Fuerst was responsible for the project and for offering assistance to the students when they needed it.

Maria's role was to implement SSEP in the division, train students and teachers how to compete in the program and answer questions that Leslie needed clarification about during the project. If

Maria did not have an answer, she consulted with Dr. Goldstein, Director of SSEP. Maria would inform Leslie and Leslie the students. It was a team effort to help the students be successful. It was Leslie who guided the students on their project. The students did ninety-five percent of the project on their own.

Maria did call them "my kids" because this is what she calls any student she teaches, directly or indirectly. The media on occasion would assume that the students were Maria's. Maria spent a great deal of time with the students and Ms. Nesbitt Fuerst including the press conferences, packing the experiment, the trip to Washington, the launch party at the Manitoba Museum, the landing of the experiment and retrieval, getting the testing set up at CancerCare, and the AAIMs Day Royal Couple presentation. As Maria stated in an email to me, "In a sense, I do call them my kids as I am so very proud of them."

It is important to note Maria's dedication to children. She had students at her school in hers and Mr. Enn's classes, and students from the Brant Argyle School in the Interlake District, all chosen to go to Capital Heights, Maryland. The Brant Argyle students were having some problems with their experiment. On Facebook Maria wrote:

•   Oct. 20, 2012 "I was able to salvage another school's dying experiment (was close to not being done, due to technical & biological constraints) I was able to work with an amazing team at Cancer Care to problem solve for Ms. Nesbitt Fuerst and her group of kids competing in the Student Spaceflight Experiment Program, YES ! There is going to be one very happy teacher mentor and her group of kids who were losing hope about their entry!"

•   Oct. 22, 2012 "165 days until launch of Interlake kids experiment to the space station. Had one VERY VERY HAPPY teacher & group of kids today, when I told them I was able to salvage their experiment."

Maria's selfless determination to save the Brant Argyle experiment is commendable. If this experiment could not have been saved, the experiment from her class on Royal Jelly would have gone to the ISS instead.

***

A more detailed explanation of the Argyle students' experiment was provided by the Manitoba Institute of Cell Biology, Nov, 2012: "Liz Henson and Eilean McKenzie-Matwiy established a new collaboration with the International Space Station and Argyle School. With expert help from Dr. Dan Gietz (UofM BMG) and direction from Ms Leslie Nesbitt Fuerst and her Grade 5 students Ryan, Avery and Ethan, the students came up with an experiment where samples of yeast will be sent to the International Space Station on April 5, 2013.One yeast sample will be suspended in saline and another sample in green tea made in saline. Since the space station is exposed to higher levels of cosmic radiation than on Earth, the Grade 5 students will measure whether yeast undergo DNA damage and whether this DNA damage is protected by antioxidants in green tea by scavenging free radicals. This project will teach the students about preventing cancers. There were 1,254 entries from across North America and Argyle's is the only Canadian entry that won. The commander of the space station will be Chris Hadfield, the first time a Canadian takes the helm of the ISS. Canadian yeast in space for cancer research is truly remarkable. We are proud of the students and to be part of this project. Congratulations!"

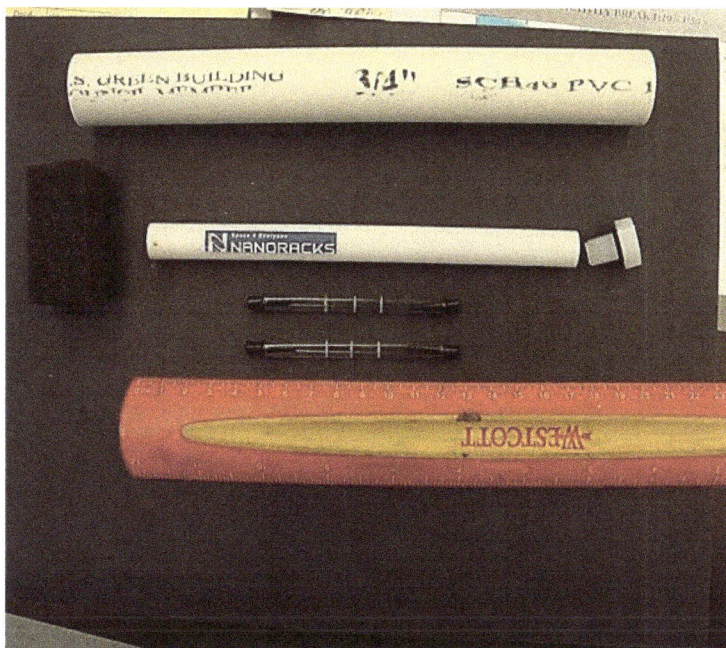

This is a TYPE 2ME (Fluid Mixing enclosure, left). Students put an item in the smaller tube with the blue sticker on it and the one small glass ampoule can have a substance inside that will be cracked by the astronaut and released into the white tube to mix with the substance in there. The astronaut has the opportunity to shake it to mix it for the students based on the instructions given by the students. Due to a failure of this system in Mission 2, a new set was developed and midway through the students had to learn how to use this one for their launch to the ISS.

***

The Interlake School Division is the first community in Canada to participate in the program. Throughout the 10 week program, students worked with scientists to design and propose experiments to fly on the International Space Station. Two hundred proposals were submitted to the program throughout the division and three finalists were chosen, including Nickel's class. goo.gl/4g5Kpo

"69 days until launch of SSEP Interlake kids cancer experiment. Things are moving fast now: March 16th we have to have our experiment in Nanoracks hands in Houston for launch loading. GETTING SO REAL NOW!! launch location is not where our kids can go, may be Kazakstan!! a tad bummed:(but we will try to send them to the conference July 2-4th in Washington DC. We are going to host a HUGE LAUNCH PARTY INSTEAD, DETAILS A COMING SOON!! GO CANADA" (FB Jan 25, 2013).

"2 more sleeps until my Space Knights Club kids put on their space science riot activities for the whole school, honoring Chris Hadfield and our SSEP experiment to the ISS" (FB Feb 26, 2013).

"Finishing up last minute touch ups for space riot (had a great group of space club gals also help me out) , now to run to Walmart I need glue sticks, and ice my astronaut cake (3-D) , get out my flight suit. Tomorrow busy day. We have CTV news at 6 pm coming, Argus and Tribune coming, and live radio interview at noon on space riot and our experiment to ISS. We are go for launch!!" (FB Feb 28, 2013).

"Just got a very nice email about a posting on the National Student Spaceflight Experiment Program that my school division is in and I direct. GO CANADA BE PRPID [PREPARED] WE HAVE AMAZING KIDS IN MANITOBA This post is dedicated to SSEP Mission 3 community Stonewall, Manitoba, Canada, whose student flight team — current 5th graders and future Doctors Enns, Good, and Petricig are awaiting launch of their flight experiment on SpaceX-3 Dragon out of Cape Canaveral Air Force Station, Florida.

However we were delayed yet again and did not launch in April or May 2013 due to the denial of the experiments biologicals by the Russians, so we were moved to a June 17, 2013 launch date on the Space X Dragon. That was delayed again due to a faulty rocket engine and we were moved to Sept. 7, 2013, then delayed to Sept. 25th, 2013. That got bumped again to Nov. 7, 2013, then again to Dec. 17th, 2013. While we were in Washington a major ISS emergency happened with the cooling of the ISS, so the launch was cancelled due to the astronauts having to do an emergency spacewalk to save the station, it was a very serious situation. We had to explain a great teachable moment to the students about it, similar to Apollo 13 seriousness. So it was delayed till January 7th, 2014. We did go to the launch site and saw where our experiment was housed and spoke with NASA & Magellan engineers about it. We came home and set up a launch party with all the students from Brant Argyle Elementary and a few chosen reps. from my space club to watch the launch live at the Manitoba Museum of Man & Nature take off successfully on January 9th, 2014. The experiment returned to Earth in Kazakhstan, Russia and we received it March 13, 2014, but official opening of it took place on March 17, 2014 at Winnipeg's Magellan Aerospace Ltd. You may view the latest post at http://goo.gl/41sYYh (FB May 17, 2013).

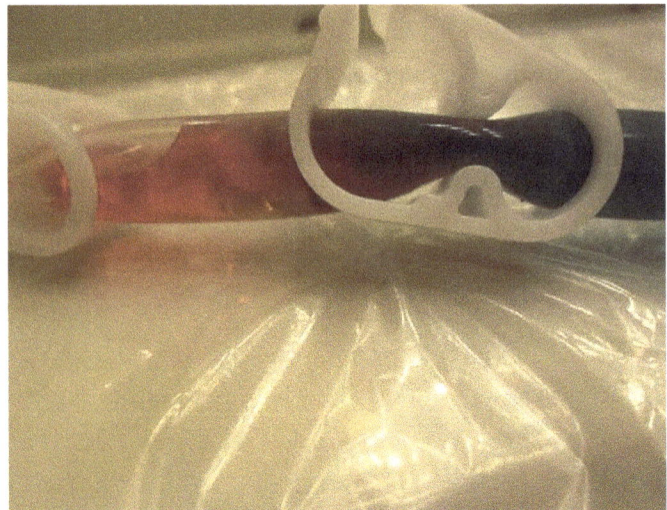

Picture (right) of the new Fluid Mixture Enclosure for our experiment aboard the ISS

"I am So honored and thrilled to be keynote speaker at the EdCamp Winnipeg conference tomorrow morning from 8:30 - noon, to talk all things space and my space club kids and our experiment to the ISS. excitedly nervous. Then back to prep for baby shower, going to be awesome" (FB May 31, 2013).

"So excited, today is my last day for the year to wear my flight suit. I present my year end and grad certificates to my space club kids in front of the whole school and show our power point of pictures of all the fun we had all year with them. We are Go for launch!!" (FB June 17, 2013).

"Got some wonderful news today!! Been asked to be a guest speaker for Manitoba Aerospace Week kick off Sept 10th, and speak at a meeting with the Manitoba Aerospace Assoc. and help Magellan Aerospace launch their exhibition and permanent display of the Brandt rocket used in space experiments around the world at the Manitoba Museum and they want my space club kids to come as well!! Way cool! And I got an email from a physics teacher in BC interested in the Student Spaceflight Experiment Program and wants to know more, he has been following my story on the national news. so cool!!" (FB June 19, 2013).

"Great planning mtg [meeting] with some awesome Aerospace movers and shakers in our province. This event is going to be awesome for my space club kids, our experiment group and our province. Lots to do over the summer for this, have to be prepared well for fall. Don't want to screw this up, since I play a key role in the fall. Can't say much more , but details coming 2nd week in August. That is when I can say!!!!" (FB July 15, 2013).

"OK OK OK!! I can finally share my big news coming up!! I have been working with Manitoba Aerospace , Museum and Boeing in all the planning this summer for Tuesday Sept. 10th. That is the day I have been asked to speak at the Annual General Meeting & be the keynote address at their breakfast mtg at 7:30 am (yes, that is in the morning) and then they have invited my space club kids to come down to the Museum after the mtg to watch the opening of the Black Brant (rocket) display at the museum. The Black Brant is a Canadian-designed sounding rocket built by Bristol Aerospace in Winnipeg, Manitoba. Over 800 Black Brants of various versions have been launched since they were first produced in 1961, and the type remains one of the most popular sounding rockets ever built. They have been repeatedly used by the Canadian Space Agency and NASA for suborbital research." (FB Aug 25, 2013).

"The media are coming, Government Minister of Entrepreneurship Training & Trade (our main sponsor for the ISS experiment) is going to make an official proclamation and ribbon cutting ceremony, and our ISS experiment kids are going to be there (1st, 2nd & 3rd place) to talk about their experiments and to highlight our 1st place winners and the upcoming launch Nov. 11 to the ISS on board the Space X Dragon. I am so so so so proud of my kids and what we are going to do on the 10th. And after that we are going to take them to the new planetarium show there as a treat!" (FB Aug 25, 2013).

"Super super nervous, BIG BIG DAY for my space club kids, Student Spaceflight Experiment Program Winners 1st, 2nd, & 3rd place winners who are presenting their experiments to a whole wacka media outlets and our Provincial Aerospace AGM mtg at the Manitoba Museum to kick off the launch of the Black Bryant rocket display by Magellan & the minister. Super excited but nervous. My girls are so excited to present, they go so into it again and they acted like they won the experimental spot and their boards look like it too. So super proud of them all. Humbled to be their teacher and mentor, they amayze me!!" (FB Sept 9, 2013).

"Houston we have success!! Great great day today. Networked with some movers & shakers in the aerospace industry and even had a lady in charge of National Defence come up and ask if we would happen to want to have Ooooooooo Newest Cdn Astronaut Jeremy Hanson come to my school to do a workshop with my space club & visit. Had so many complements on my presentation about how inspiring it was to them and the work I do. I felt really good & they loved how enthusiastic my kids were and how stunned they were at their type of projects. The minister of Entrepreneurship Training & Trade came by to see me personally and all of my kids. Minister Peter Bjornson a former teacher. was awesome and I got new contacts for getting funds for our launch prep to see the launch in Dec. a fantastic day MY KIDS ROCK!!" (FB Sept 10, 2013).

"Business networking mtgs all day today with all the major players in MB Aerospace Industry (a little intimidated, lil o teacher in their midst) trying to secure launch funds for our ISS experimental grp winner and hopefully the conference in July at the Smithsonian. Oh and MONDAY is launch #1 to the ISS, our mission patches go FINALLY, keeping fingers crossed it is a go. BIG presentation to our winners Monday!! busy busy today (Sept 12, 2013).

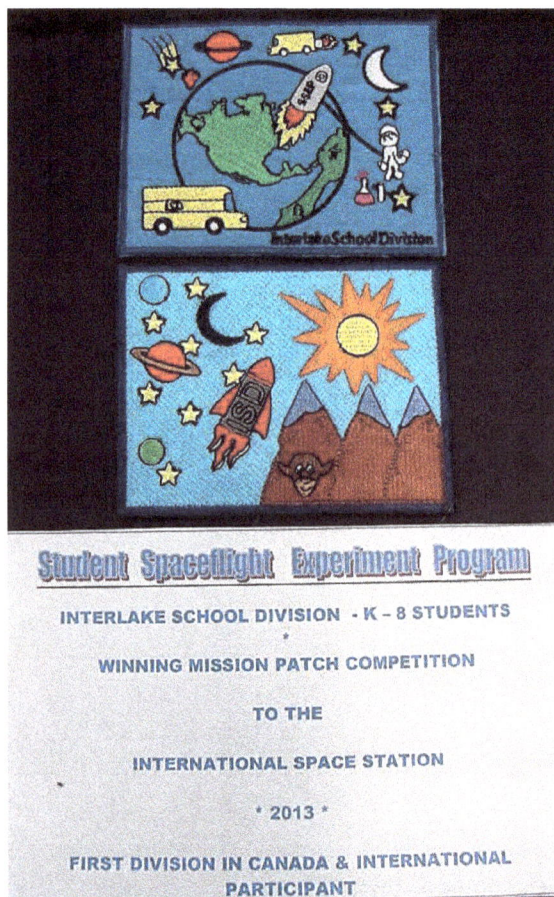

Student Spaceflight Experiment Program

INTERLAKE SCHOOL DIVISION - K - 8 STUDENTS
*
WINNING MISSION PATCH COMPETITION

TO THE

INTERNATIONAL SPACE STATION

* 2013 *

FIRST DIVISION IN CANADA & INTERNATIONAL PARTICIPANT

"Made the Winnipeg free press special insert section kicking off Manitoba Aerospace week Sept 10 - 13, 2013. They did an article on our experiment to the ISS and my role in all of this. http://publications.winnipegfreepress.com/i/16958 1" p. 10 (FB Sept 12, 2013).

"We are go for launch Monday as long as the weather holds. The Cygnus launches on Monday around 10 am our time for the International Space Station with our 2 mission patches on board. Unveiling of the patches that have been embroidered will be Monday at board office at 10 am, so excited, all the winners get a copy. Just sent out a press release , I hope they come to see it, would be nice for the kids" (FB Sept 13, 2013).

Hi there everyone!! Been getting a lot of questions surrounding why our mission 3 had to be split into 2 launches .
(Photo credit: NASA/Aubrey Gemignani)

Mission 3 Historical Notes – Why Designation of Two Separate Mission 3 Experiment Payloads here is the write up on why : goo.gl/UxAAtc (FB Sept 14, 2013).

"… our payload of some Mission 3 experiments and our 2 mission patches from Interlake School Div Canada is on board, it launches at 8 am our time xoxoxoxoxox (FB Sept 15, 2013).

Orbital's ISS cargo delivery demonstration launch from Wallops Island, VA has been delayed one day to September 18, 2013. More information at http://goo.gl/41sYYh (FB Sept 15, 2013)

"We are go for launch at 8:50 am Winnipeg time!! GO Interlake kids mission patches and all of Mission 3's patches. Go Canada!! I am going to be a mess when the launch comes. I just can't contain myself!!/o/o/o our kids rock!" (FB Sept 16, 2013).

"Well we are delayed AGAIN!! but only 2 days, we had some media come out (newspapers) we did hand out our mission patches embroidered, lotsa parents came out, and superintendent and a trustee, so cool,kids were super proud and excited to see it in a crest format):):" (FB Sept 16, 2013).

 "HISTORY MADE TODAY FOR Canadian kids elementary education. INTERLAKE SCHOOL DIV kids 2 mission patches launched at 8:50 am on the historic first launch of the CYGNUS transport . watch goo.gl/9jkyDn (FB Sept 18, 2013).

"Here are the 2 winning mission patches (previous page, Gr 4 and Gr 8) that will dock with the ISS on the Cygnus at 7:58 EST * on Tuesday Sept 23, 2013 [previous page] , I gave all winners, all schools involved, and our school board office a copy as well. LOVE IT" (FB Sept 23, 2013)

The mission patches were sent with Mission 3A experiments, our transport was Mission 3B. Due to all the changes and hiccups with the refrigeration of experiments, we had to split up the mission. So our Canadian experiment went as Mission 3B with Mission 4 experiments.

" "OMG OMG OMG OMG, The Canadian Space Agency retweeted one of my tweets today about our patches on the ISS today, ahhhhhhhhhhh I have just made the oscars for space!! where is my red carpet dress..... my flight suit" (FB Sept 29, 2013).

See a cool video of the Canadarm II releasing the Cygnus Space Capsule in this time-lapse video goo.gl/Yx6WsC

"Ok this is getting so very real now! Fed Ex shipment to come to school for 5pm and I have to be there to sign for it and collect the prepped FME (the unit that will house our experiment). Then mtgs with Magellan Aerospace at the end of the month for securing the device in our container and working out logics. Then have to set up dates and times to fill & pack experiment for transport to Magellan by Nov 8th and then shipping to USA NanoRacks, in Houson, TX by 21st of Nov. Then we go to launch Dec 8th at MARS facility Wallops Island , Virginia. OMG this is getting fo' real" (FB Oct 7, 2013).

"Got our pkg from NamoRacks (right) for our space station elem expt to be house in. They are gas infused to prevent them from collapsing or exploding" (FB Oct 9, 2013).

"Commander Nickel is busy with NASA flight safety review on Experimental details confirmation for Nanoracks & NASA flight safety review & Flight Configuration details. This is so so so cool! I am so loving my job right now! I wish I did this full time!! Deadlines for sign off fast approaching along with prepping and shipping of the experiment on Nov 8th. Things are a moving fast now" (FB Oct 23, 2013).

"Love it when things fall into place for a change, I love my superintendent and assistant superintendent, so so so so wonderfully supportive. Space Station Experiment is going to be happening in Dec, no final lock down date but it will be after Dec 8th, still waiting to hear from NASA on final consolidated launch dates & times. We are a go!!!" (FB Oct 24, 2013).

"I rocked it today with science meetings , NASA flight safety review & flight configuration, CancerCare Liz you are amazingly inspirational, the kids are so lucky to have you in their corner. Fabulous team meeting at MAGELLAN AEROSPACE Wpg [Winnipeg] to discuss media blitz , shipping logistics, crate set up, refrigeration, area setup, location for kids entrance (originally kids were not going to come, but I discussed & negotiated for them to be present, fill and close the crate for shipping to NANORACKS , Houston and then on to NASA Wallops Islands , VA MARS facility. All personnel involved : Phil, George, & Karleigh were stupendous to work with. LOVE it when a team actually works together to make something a reality and will do what it takes to make it so!! Still working on funding to

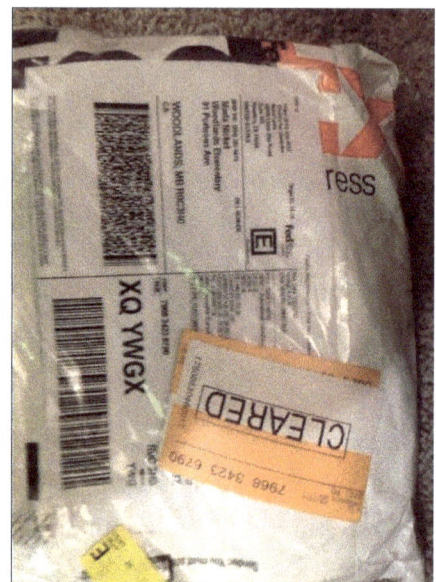

get the kids to the launch, we are a long ways to go!! I need to win the lotto /o/. I want these kids to see this launch so badly after all the hard work they have done and all the delays (we are sitting on 6 delays for our crew). It will happen, I believe it will and by golly I am going to make it happen!" (FB Oct 25, 2013),

"Our official Nanoracks and NASA approved Fluid Mixing Enclosures for filling our Cancer expt to the Space Station . I especially like the note one from NanoRacks saying 'Have a great flight.' So wicked cool I get to do this and proud to be involved!" (FB Oct 26, 2013).

"Some good news, got an $800 donation towards flight for one of my students for our trip to the launch, now have a grand total of $1100. Very shy of the $10000 needed for all the costs with it. Keep plugging Nickel, but it is tough, lotta no's and very few yes's. & I am burnt out already, and it is not even June yet!! oh dear, something is going to give" (FB Oct 30, 2013).

"Today is a big day , pre-flight experiment prep with the winning group of gr 5 kids (who are now gr 6) ,yes that is how long it has taken us to get to flight status on this ISS experiment. So excited" (FB Nov. 5, 2013)

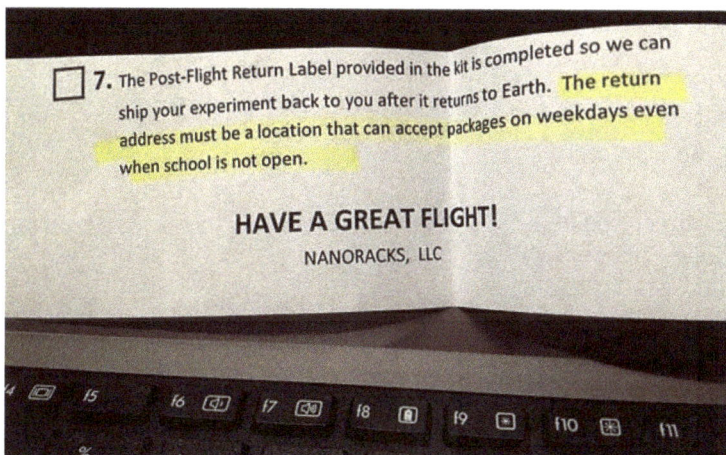

"Today's prep went very very well, the kids were so nervous but came through with flying colors, we are ready for Friday's final of loading the experiment. Today was so so real, now we are that much closer to launch status of Dec. 15th" (FB Nov. 5, 2013).

"Busy putting together my flight checklist like NASA does and what I did at Advanced Space Academy!! This is so real now, 1.5 yrs worth of hard work is coming together, JUST DON"T FORGET THE FME's to load and NANORACKS PAPERWORK /o/o/" (FB Nov 7, 2013).

"OK just got word that our Dec 15th launch time is going to be no earlier than 1:00 pm EST!!! at Wallops Island , VA MARS facility!!" (FB Nov 7, 2013).

"Last minute pre-flight safety check of materials: flight safety documentation, check, all pre-treated gas infused FME's (fluid mixing enclosures), check, Chris Hadfields mission patch for good luck, check, my squeeze rocket from NASA JSC Houston for kids to relieve stress, check, signed pictures of 3 of our top Canadian Astronauts: Robert Thirsk, Julie Payette & Chris Hadfield for the kids to see where they may be in the future, check, my advanced space academy backpack with NASA Ed zip folder to keep all the papers & pics in, with the shuttle mission patch on it for where I came from on this journey of the last 6 years and to put everything in, check, granola bar for later, check, camera to film and record for customs and history, check my phone for Skype with Nanoracks to confer during experiment prep, check, all phone numbers for media on standby, check, and finally my flight suit, check!! We are go for experiment prep launch!! GO INTERLAKE KIDS & DIVISION, SSEP, NANORACKS & CYGNUS !! Now to shower…(FB Nov 8, 2013).

"OMG we made MSN CANADA front page news , yay for our kids" (FB Nov 8, 2013). See the video about the three students experiment that is going to the ISS at http://goo.gl/saeAqZ

THE EAGLE HAS LANDED! Interlake expt landed at 9:44 am at @NanoRacks and cleared customs no problems. So we are closer to launch now!! YES. Moi is so so so very relieved now" (FB Nov 21, 2013).

"14 days till Interlake experiment launch to ISS!! OMG CLOSER WE ARE!! The Canadian Consulate has been invited!! SO SO COOL! We meet the ambassador Dec 16th day before launch!! Private meeting & photo op on balcony, apparently best pics to take in Washington!! Now do I wear my flight suit or do I wear nice clothes??? hmmmm thoughts space camp buddies? and friends" (FB Dec 3, 2013).

"Its official, here is the movie trailer our kids have done about their experiment to the space station . launches Dec 17, 2013 at 10:05 pm EST NASA TV LIVE will carry it. let's get this thing vira goo.gl/sKxk3Hl: (FB Dec 6, 2013).

Students doing their experiment with CTV-TV videotaping (right). Photo courtesy of Stonewall Argus and Teulon Times.

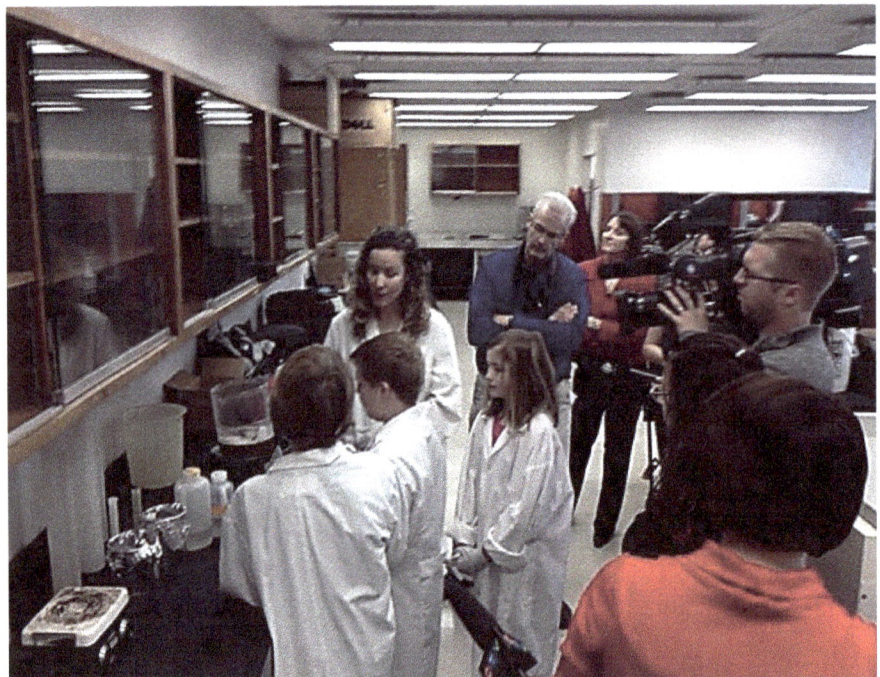

"Ok now nervous! just got word NASA launch delayed again to Dec 18th , now I have to run around and make changes to tickets and hotels etc" (FB Dec 9, 2013).

"Our little movie trailer made the National SSEP website for all across USA and the world to see. Thanks Dr. Jeff Goldstein!! goo.gl/nxGXFX (FB Dec 10, 2013).

"Launch date has changed again now to Dec 18th! forcing us to do cancellation fee changes of $150/person plus the change in airplane txs & increased hotel costs which is adding another $4500 in total (the $150 is covered off with insurance, but the ticket increase which is another $300/person) Oh man I need a Christmas miracle now!! Santa are you listening /o/. Now I have to call Cdn Embassy to inform of changes, hotels to rebook, cars, and itinerary changes to all this and inform the parents so they can rebook insurances and get me the money, cause all this had to go on my credit card,was on phone from 6 - 11:30 pm last night with all this (and I'm still on phones about it today) . eyyyiiiyiiii. Santa I been really good this year, hope you are listening! /o/o" (FB Dec 10, 2013).

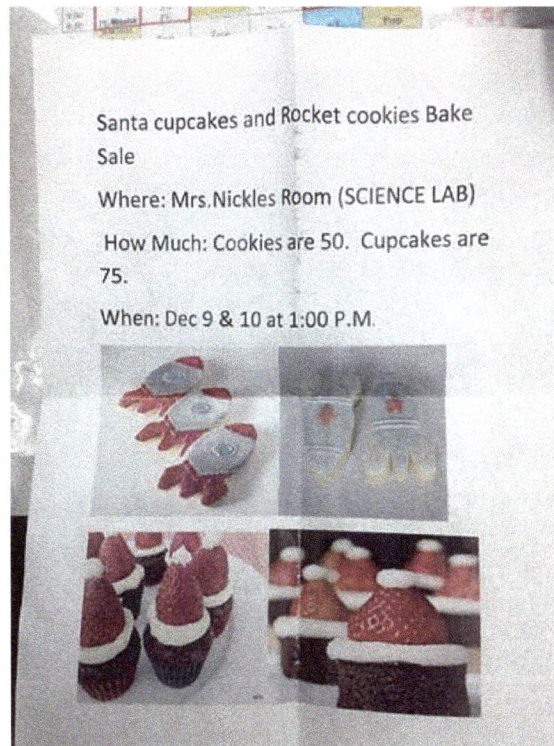

Students fund-raising poster that was the start of the flood of money donated for the trip to see the launch at Wallops Island, VA

"In amidst all the space craziness , late nights last few says over all the conflicts ,confusion & stress. Yesterday, my space club kids all sold their cupcakes and cookies and made $70 to give to the experiment kids. To top that off, one of my grade six girls comes up and says Mrs. Nickel here some more money " I don't really need it, but you do. Here's some help for you it was $6 from her own money. I broke down in tears over her thoughtfulness, I got great students!!" (FB Dec 12, 2013).

"Had the most unexpected call today from a random stranger in MB who called the school and asked for me & said he was retired and heard of our troubles & said he was donating $2000.00 of his own money to our plight. I got so emotional on the phone with him and said I could hug him & thank you. Then at 5:30 pm get an email from the Winnipeg Airport Authority that they are donating $2000.00 to our launch cause. That now covers our flight issues, now just $2200 in extra hotel costs and we are saved by our guardian angels" (FB Dec 12, 2013).

"Ok, today has been the MOTHER OF ALL DAYS OF THE CENTURY!!!I have been in crisis mode since Monday, am running on 3 – 4 hrs sleep a nite since Monday. Plus two xmas concert productions at school to help with and deal with my spaceflight crisis. If it was not bad enough , having to change flight & hotel dates and the costs and being short $, today has been my worst night mare come true. I have been juggling 3 different phones all day, conference calls, emails up the wazoo with short massive timelines to address and coordinate and be ready for. The ISS crisis currently is serious and affecting our launch status. Was told we are strongly getting word that we are all (all SSEP 3b & 4 missions) possibly are cancelled and moved to mid to late Jan due to ISS crisis (waiting to see if our transport is going to be used as an emergency repair parts supply mission to assist astronauts in the spacewalks to save the cooling leak on Station). I have had to coordinate with our scientist researchers at Cancer Care the possibility of a quick turn around to get a new mix stick to us (ours would be shipped back), had to let them know by 4 pm today our decision, only got word on this part at 10:30 am. talk to see if it could be done, have run into the problem of University shut down and cancer care research area from Dec 20 to the 2nd of Jan. putting strains on our kids and their ability to get it done (a whole set of other logistical issues to that) , coordinating with Magellan to see if we can ship it back and use customs overnight or that day to get it back to launch site in time for vehicle integration with the Antares rocket, they said yes, discussing possibility of a cancelling of launch and what needs to be done for that and all the paper work needed for all 11 people going for insurance claims , been putting together pkgs for that since yesterday, needing superintendent letters for insurance purposes for today to go out , conference calls with space systems engineers about the prospects that our launch will be cancelled and moved to Jan. , then it was also trying to teach in the middle of all that and all my kids (knowing the ISS crisis and our launch issues, all of them today from Gr 5- 8, were " Mrs nickel, what do you need us to do so you can get that experiment back to space ?" before i could say anything, " they all said " OK everyone, get your work from the table help each other out, and quiet with the phone rings" right after that my phone rang and the room went deadly silent, with all eyes on me, ears straining to hear, and super quiet whispers about what was happening and voices saying" guys be quiet, Commander Nickel is saving our experiment, let her work in quiet, just do it!"" and they were so amazing! then kids asking about the call and questions on what is happening and concerned looks on all their faces. Super cute! Then secretary & EA's watching my classes for short periods for phone calls to me from my staff room about SSEP. Dealing with sponsors calls to help with funding our plight to many , many Phone calls from the winning school and their admin to coordinate strategies and contingency plans for if we are cancelled before Sunday. More calls & emails, no snack or lunch today due to all demands. Making sure all lesson plans are ready for next week and all that work is a go. Realized I was hungry on my way home at 6:45 pm and hubby calling me in the car to say to get sushi he ordered for us , place is on wasy home. WE are in a waiting pattern, be advised Monday Dec 16th (we fly on Sunday) NASA makes the call, based on ISS space walks in the next few days. I hope we get that before we leave. really want kids to see launch and so do I, have never seen a live launch.

Today and all week have felt like I have been producing the JUNOs like you did when they were here Brenda Bourns , I am living on adrenalin right now!! oops here goes another call, gotta run......." (FB Dec 12, 2013).

For Maria, "Failure Is Not An Option."

"I would not have made it this far without the wonderful support of amazing space ladies , so many friends & family near & far: Diana Bosley LeSueur, Becky Loy, Jacqui Flowers, Marian Gilmore, Jennifer Cheesman, Jennifer Kelley, Kaci Pilcher Heins, Cindy Sigurdson Brad, Monika Deschutter, Brenda Bourns, Dawn Sirdar, John Stangroom, Teri Slobodzian, Tracy Marchak, Lisa Marquardson, Tricia Orsulak, Shelley Miller, Stacey Nickel, Susan Cowan, Mary Chalmers, Bea, Debbie Gajdosik, Kerri Jo Nickel, Lindsey Kristin, Susan George and I know there are more that have posted such kind words of encouragement. That is worth so much to me, even if there was nothing concrete you could do for me, this is so very much loved and helpful in so many ways, I TRUELY THANK YOU FROM THE BOTTOM OF MY HEART XOXX" (FB Dec 13, 2013).

"I AM SO CRAZY HAPPY EXCITED RIGHT NOW!! I want to thank the efforts of MB Aerospace Director Ken Webb for all his help and shout outs to the MB Aerospace community, Terri Trupp from Boeing , PAscal Belanger from Wpg Airport Authority for stepping up and fully funding all our needs right now! WE DID IT, ! & THANKS TO CTV 'S BETH MACDONALD & JON HENDRICKS FOR GETTING THE WORD OUT IN THEIR STORY. I AM BEYOND HAPPY, WE ARE GO FOR LAUNCH ON THE 18TH STILL." (FB Dec 13, 2013).

"Mission Control, we are a go for launch T - 5 days and counting! Canada is a go for launch, fly!! My flight director skills training & being assigned as commander of the ISS simulator from Advanced & regular Space Academy have proven immensely & powerfully helpful, helped me to navigate this real life NASA / SSEP situation all this week and since June 2012 when I first started my journey. Thank you Marian Gilmore for giving me that change to be selected and a part of the wonderful Honeywell Educators family!! ever greatful!" (FB Dec 13, 2014).

"OK ! NOW I really feel like GENE KRANTZ now!! I am in the missions management loop with the Canadian Space Agency on this crisis on ISS & with our launch updates with Orb-1 and the Cygnus!! Let's do this!!! Off to all final mission preps and errand critical to our success down them, wish me luck!!" (FB Dec 14, 2013).

"we are delayed in launch to Dec 19th, they load cargo once NASA gives the OK tomorrow, it was supposed to be done Sat, but will opt for Sunday once NASA gives green light. At NASA's direction, Orbital's Cygnus operations team deferred loading the mission's final cargo into the spacecraft earlier today, postponing that operation by a day. Orbital will await NASA's direction for the final cargo load tomorrow while the cooling loop issue aboard the ISS is being investigated. If we get the go-ahead to load

View from the Canadian Embassy (right). Maria and students are in Washington, D.C. (FB Dec 16, 2013).

the final, time sensitive cargo on Sunday, roll out to the launch pad would be on Tuesday, December 17, launch on December 19, and rendezvous and berthing with the ISS on December 22." (FB Dec 14, 2013).

"time for bed, 2:00 am is going to come early for me and 2:30 taxi pick up to airport, but I can't sleep I am SO STUPID EXCITED RIGHT NOW!! I CAN'T SLEEP!! WHO CAN SLEEP KNOWING CANADA IS ABOUT TO MAKE ELEMENTARY HISTORY!! GOING TO THE ISS" (FB Dec 14, 2013).

"Update on our Cygnus ISS launch ! Time change! It will be still Thursday Dec 19th but at 9:19pm EST & NASA TV live will start broadcast at 8:45pm EST with chats about what is going up & why! A sort of play by play until blast off! We r still a go! Washington is way warmer than wpg I'm so surprised! I drove the big 11 person van with all inside did pretty goods. Made it to hotel. We see ambassador tmrw flight suit here we come!" (FB Dec 15, 2013).

"A very big special day tomorrow, I have arranged a HUGE surprise from the Canadian Space Agency with Bill Macky CSA rep who works at the Canadian Embassy, he worked at NASA for a time in Mission Control Houston and the robotics interfaces & with Chris Hadfield, as well I am in talks with NASA HQ in Washington DC, to meet at the Smithsonian National Air & Space Museum at 12:30 pm for all the fun surprises for them & the parents. They just know tomorrow

they are going to the museum after we eat breakie and have a session at 12:30 pm I can't wait to see their faces when they see the shwag they are going to get! I AM DA BOMB!! & PROUD OF WHAT I JUST PULLED OFF, & it will be an even bigger coup if I get them on the launch pad tour that the other USA SSEP get to do!! I have my "people" working on it, we were told that we could not because we were foreign nationals, BUT I got my embassay & CSA people on it, no promises BUT I REALLY HOPE I CAN PULL THIS OFF!! wish me luck !!" (FB Dec 16, 2013).

"We [Orbital Sciences Corporation] have been authorized to load the final cargo for the International Space Station into the Cygnus spacecraft today. This cargo is about 95kg of time-sensitive materials consisting primarily of science payloads. We expect to finish loading the cargo later this afternoon. Installation of the Antares payload fairing is scheduled to occur on Monday."

"Just got word that NASA had scrubbed our launch officially! 3 emergency space walks to fix coolant issue! So it will be mid to late Jan. For new launch of our expts. I'm totally bummed out on this kids are too." (FB Dec 17, 2013).

"A special meet & greet today set up by myself, Bill Mackey from Cdn Space agency & NASA program ed. Kids met its deputy assoc admin of education programs integration , James L Stofan who brought NASA mini backpacks with NASA goodies inside [Smithsonian Air and Space Museum]. A big highlight! He was amazing with the kids! A fascinating man, he used to be a trainer of Shamu the whale at Seaworld!" (FB Dec 17, 2013).

"We are going to Wallops Island, VA MARS facility [the intended launch site for their experiment] anyways ! I have activities set up for the kids with Magellan & NASA reps to have closure and still see the rocket launch pad! We r going to make the best of things." (FB Dec 18, 2013).

Today we go to Dulles Airport to see the Shuttle , I am so crazy excited to see her again and think of the moment I touched her wings at the VAB [Vehicle Assembly Building) at NASA's Kennedy Space Center as well as see Discovery's engines being taken out for decommissioning

(I SAW NOT 1 BUT 2 SHUTTLES THAT CLOSE) on our tours there with Advanced Space Academy, I may not have ever seen a launch but I got to see her THAT CLOSE AND BE HER PILOT IN THE SIMS we did, I will always treasure that forever! GOOD DAY TODAY, YESTERDAY WAS SO AWESOME TOO," (FB Dec 20, 2013).

"Just got word our experiment will be heading to the space station with Orbital on a new launch date of May 13, 2013. I'll need some advice on hotels, airports, etc." (FB Dec 20, 2013).

On December 21, 2013, Maria, the three Argyle students and parents got to see the Lincoln Memorial (right), Washington Monument & the White House. It was a gorgeous day with a temperature of 15 degrees C.

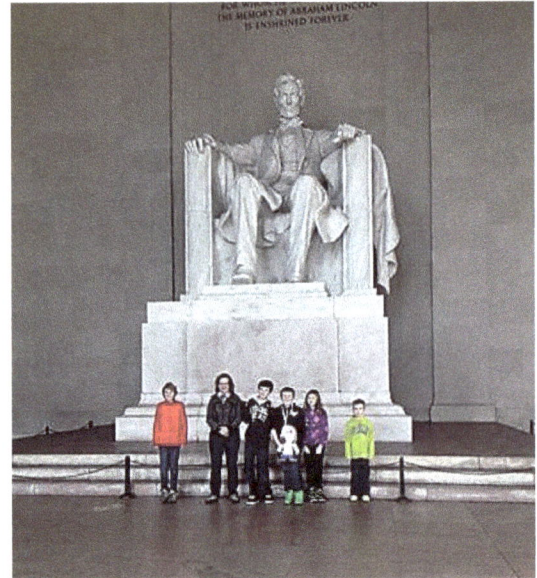

"Hoping weather is fine in Ottawa tmrw so we get home tmrw and not have to overnite it in Ottawa! This has been a challenging week here in Washington & Wallops been working non stop & juggling Parents, kids , events, ambassador, NASA, CSA & SSEP ! I am so looking fwd to be home to slow down & soak this trip in! It was a lot! I met so many new contacts & friends in space for life out here very grateful & NASA den is going to do something for my space club kids once back I am super excited & happy to have new things for my always greatfull space club kids who accept anything. at anytime ! Miss them lots! Love em forever! FB Dec 21, 2013).

"Got a tweet from the Canadian Space Agency that they favourited & retweeted my pic of our expt kids in front of the Canada Arm on Dislplay next to Shuttle Discovery at Dulles!! Cool" (FB Dec. 23, 2013).

Launch vehicle for experiments and astronauts supplies (Photo credit: Orbital Image Corp)

Launch vehicle for experiments and astronauts supplies (Photo credit: Orbital Image Corp)

We are go for launch now Wed. Jan 8, 2014 at Manitoba Museum in the auditorium near the science gallery at 11:30 am. Just got some amazing pics via our awesome NASA Wallops engineer Catherine, taken Dec 16th when we were there and they were prepping her for launch that is now happening in 2 days. Still cool pics (FB Jan 6, 2014).

"Launch cancelled again! space radiation reason, launches tomorrow at 12:10 pm wpg time or 1:10pm EST" (FB Jan 8, 2014).

"We are still a go for launch tomorrow [January 9], a bit earlier at 1:07 EST which is 12:07 Winnipeg time!! PLEASE TAKE OFF TOMORROW!! WE HAVE BEEN BUMPED & DELAYED AND WAITING SINCE APRIL 5TH, 2013

"We are a go for launch ! (Image below: Orbital Science Corp). The rocket propellant is being loaded right now! Ctv , CBC TV CBC radio & CJOB radio all here to do story so exciting! And Canadian space agency just retweeted us on launch today" (FB Jan 9, 2014). (Photo credit: NASA/Aubrey Gemignani)

That WOW moment for Ethen Enns, Ryan Petricig, Avery Good and Maria Nickel with the children's teacher, Leslie Nesbitt Fuerst http://goo.gl/j11BaX/

Good coverage of this Canadian first by the news media goo.gl/DRl8oV

A friend of Maria writes: "So proud of you! You have done so many wonderful things for your kids! I was just bursting with happiness for you when I saw the news - I know it was years of hard work! You are Awesome Maria!" (Connie Kosky-Levesque, Winnipeg).

Video of the launch goo.gl/yNTEtc

We made it!! Argyle kids 1st Canadian kids experiment to space is caught by Canadarm 2 today at 8:05 am EST. got a shout out from CSA on it YAY. A longer video of capture http://goo.gl/sjgaaE [Look how beautiful the Earth is in the background]. A shorter 2 minute video showing the docking of Cygnus to the ISS. goo.gl/wUQdsj

"OMG!! I was just mentioned and our expt in space by newest Canadian Astronaut Jeremey Hansen!!!! and I have just been tweeting with him!! SO SO SO COOL! The students & I will be on CTV Canada AM LIVE 7:05 am on Thursday's show (Jan. 16, 2014)to talk about SSEP, Space Station & our expt currently on the ISS. So fun!! I am so thrilled right now!!" (FB Jan 13, 2014).

"ON CTV Canada AM Live with Beth tomorrow at 7:05 am Wpg time (CST) on Channel 7 & 5 ! talking about the experiment on the ISS and history for Canadian MB kids who did it! Congrats to a BC school division that is signed on to do Mission 6 to the ISS! The mvt has begun!" (FB Jan 15, 2014).

"There was a major tech glitch that should have been taken care of before we came to do this [interview], not impressed with Wpg crew on this. Test your ear pieces BEFORE you arrive , not 3 min before we are to go on air with Toronto & the adapter does not work for the student. I had to relay the info to the student Ryan, felt bad for him, missed out on one translation question, getting Terry Fox name mentioned, got lost in me relaying it to Ryan . I felt bad about that! Was choppy but it did go OK. Poor Ryan was so nervous, you could tell, and shy. But he tried his best, proud of him — feeling awesome." (FB Jan 16, 2014) See goo.gl/k6VCMD and goo.gl/iFSFF3

"Our mission patches finally arrived home from space station . They launched Sept 18, 2014 on Orb-1 Cygnus. Happy kids shortly" (FB Jan 21, 2014).

A mission patch [rough sketch right] Maria made in honour of the Argyle students' experiment to the International Space Station.

Videos of Interlake Experiment tube being shaken on Day 1 and Day 2 goo.gl/t1RQiD

"I am so blessed to have such a wonderful crew of space club kids, these guys are the best crew I have had so far , since I started my program!!

Now here is the best news yet!! Here is what I have been waiting on but now I have it confirmed!! Newest Canadian Astronaut Jeremy Hansen is coming to Manitoba Feb 27 & 28th. The CSA contacted me, I did not call them, they said he wanted to come and meet my space club kids , our school, the experiment kids & as many schools as we could cram into our gym. We are the only rural school he will be visiting when he is here. We get to jam with him on Feb 28th at 1:45 pm!! I AM SO WEARING MY FLIGHT SUIT & I get to intro him to all the kids & we get a private pic with our space club kids! It's going to be our final team picture for the year!! SO INSANELY EXCITED!!" (FB Feb 13, 2014).

"Got the ultimate compliment from Astronaut Jeremy Hansen. He asked me how long have u been teaching ? I said 20 yrs & he says You don't look old enough to be in it that long , you look great! I said aww thanks & you look too young to be an astronaut and we giggled together and chatted more about other items after that it was awesome" (FB Feb 28, 2014)."

CSA Astronaut Jeremy Hansen was such an incredible person to talk with one on one, several times during his stay at our school, we had such great chats about various things, was so cool to have that casual & frank of a conversation with such an astronaut. There is something very special about him, he is so generous with his time, comments , concern & passion for space

exploration and inspiring kids. You were witnessing the birth of the next amazing Canadian Astronaut & global ambassador. He is going to connect and resonate the same way our Chris Hadfield did with the world. HE JUST HAS THAT CONNECTION WITH PEOPLE WITH WORDS AND ACTIONS. I look forward to his first space walk, it will be super cool, I got goosebumps thinking of it!! & our kids got to hear from him, truly magical for them and history making .wow!" (FB Feb 28, 2014).

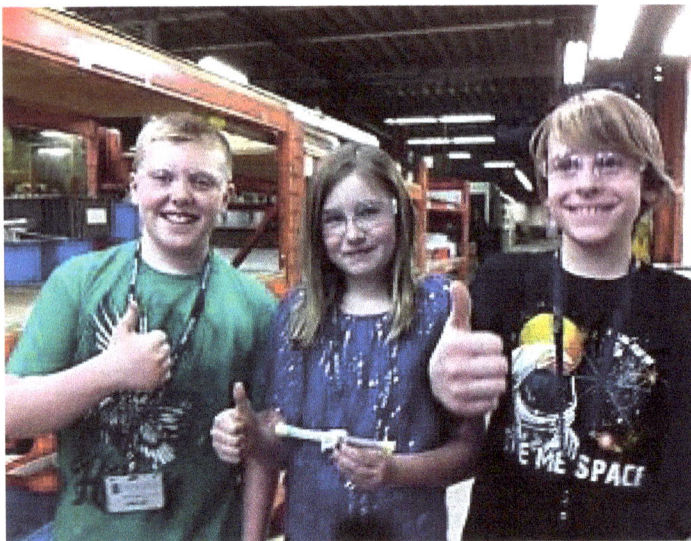

The Manitoba students' cancer experiment lands back on Earth. (Photo credit: (Photo credit: Stonewall Teulon Tribune by Natasha Tersigni)

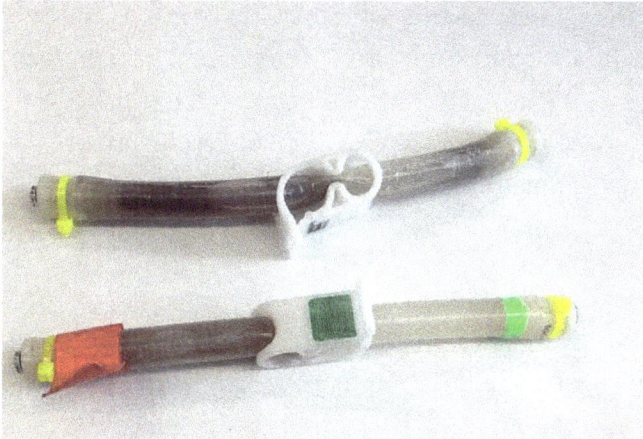

A closeup view of the experiment (left). The top one is the control or ground which means, it stayed on Earth. The bottom one is the ISS experiment.

Three Grade 6 students from Manitoba's Interlake region have received the results of a cancer research project that's been over a year and a half in the making. http://www.cbc.ca/player/News/Canada/Manitoba/ID/2442423408/" (FB March 17, 2014).

Maria, Leslie (teacher) and the children about to hand over their return experiment to staff member Elizabeth Henson at Cancer Care (right) for the culturing process that takes three weeks. (Photo courtesy of Manitoba Institute of Cell Biology - CancerCare Manitoba).

"Kinda crazy when the Canadian Space Agency is tweeting with you about your experiment back from the space station. here is today's tweet from them: CanadianSpaceAgency @csa_asc . @CommandrNickel - We can't wait to hear about the results the students' #ISS experiment! winnipeg.ctvnews.ca/mobile/kids-sp... #STEM #Winnipeg (FB Mar 18, 2014).

students'#ISS experiment! winnipeg.ctvnews.ca/mobile/kids-sp... #STEM #Winnipeg" (FB Mar 18, 2014).

***

The results of the experiment following the three weeks culturing process were as follows as written by Leslie Nesbitt Fuerst.

"Upon preliminary visual analysis of the FME, there was an unexpected color change of the FME (Earth) which led us to believe that there may be some type of contamination to the samples. This was confirmed, there was a gram positive bacteria strain which was not allowing the yeast colonies to grow. Therefore there was not enough yeast colonies to count and compare color change for the final analysis. Therefore we set up a mock experiment where we radiated samples with a projected amount of radiation that would simulate space and earth over the 10 week duration of the experiment. Yeast was regrown and radiated with 0.3 Grey (G) to simulate the amount of radiation that the Earth sample would accumulate in a 10 week period and another sample of yeast was radiated with 10G- a calculated amount that would be expected if the yeast were to be in space for 10 weeks. Although the space sample appeared to be intact, it was determined that both earth and space samples were contaminated with bacteria. The space yeast was then treated with antibiotics to allow the yeast to grow.

Speculations on how the contamination occurred are as follows: Possibly someone touched their face while preparing the ampoules, maybe the fan could have dropped dust, large number of people in the room may have created dust in the air, maybe the equipment wasn't sterile. If the experiment were repeated we would try to alleviate these errors.

Colonies of healthy and mutated yeast were then counted for roughly 25 plates for each of the following situations: Space Saline, Space Green Tea, 0.3G Saline, 0.3G Tea, 10G Saline and 10G Green Tea. An average for healthy and mutated colonies for each situation was calculated and then a percent was calculated for healthy versus mutated colonies for each situation. The percentages were graphed.

Upon analysis of the results it is evident that our hypothesis was supported; a green tea antioxidant was able to protect the yeast cells from mutation due to cosmic radiation, although the results were not as significant as expected. It was assumed that the radiation would cause much more cell mutation than what we saw in all of the samples. We were unable to determine whether cosmic radiation increased the rate of mutations in yeast's DNA due to the loss of the ground truth sample, however, our results show that there are more mutations with an increase in radiation from 0.3G to 10G, so we could speculate that there would be more mutations in space than on earth.

We thought that we would see much higher mutation rates in the space saline sample; however, the numbers were only slightly higher in the saline as opposed to the green tea. Some of our theories for this are: all the shielding on the ISS protecting astronauts from radiation decreases cell mutations, the positioning of the ISS- turning away from solar flares that cause increased radiation- another way of protecting astronauts from harm, the container that the FME was encased in as well as the FME itself increased protection- the experiment likely would not receive as much radiation as an astronaut.

If we were to do the experiment again, we would try to get a more exposed area on the ISS so that it would better replicate the environment of an astronaut while aboard. We would be very cautious in a sterile environment to ensure no contamination to the samples.

We can now provide scientific evidence that a green tea antioxidant decreases cell mutations; therefore we encourage astronauts to drink green tea, or take a green tea supplement while in space as a preventative measure to reduce their risk of cancer.

The experiment was a terrific learning experience. Now scientists are working off of our idea to expand on their knowledge of how radiation impacts cells and how to prevent cell mutations. They will be looking further into which strands of DNA were affected by the radiation. This will lead scientists to a better understanding of how to possibly prevent the mutations caused by radiation."

***

***The steps involved that Maria explained to Ms. Nesbitt Fuerst in order for the experiment to get to the ISS were:

• placing it in the FME(Fluid Mixing Enclosure) containment container designed and supplied by NanoRacks LLC in Houston, TX.

• Having a crate set up by Megellan Aerospace , Winnipeg and using their customs broker to secure the full transport to NanoRacks LLC who will then arrange for final transport and paperwork to NASA at the MARS facility at Wallops Island, Virginia for final payload loading onto the Falcon 9 Rocket and the Space X Dragon transport capsule.

The Student Spaceflight Experiment Program is undertaken by the Arthur C. Clarke Institute for Space Education (http://clarkeinstitute.org) in partnership with NanoRacks, LLC. This on-orbit educational research opportunity is enabled through NanoRacks, LLC, which is working in partnership with NASA under a Space Act Agreement as part of the utilization of the International Space Station as a National Laboratory.

Comments from Cancer Care Manitoba concerning the students experiment:

"Dr Dan Gietz, Elizabeth Henson and I feel that the students' effort was exceptional and it is a testament to Leslie and everyone at the Brant-Argyle school for instilling in them the confidence to pursue this project, patience, curiosity, and teamwork. Each student had unique skills that complemented the team and we are very honoured to have worked with such a special group young people as Ethan, Avery and Ryan." (Eilean J McKenzie-Matwiy PhD, Research Officer Manitoba Institute of Cell Biology - CancerCare Manitoba).

Maria sat down with these students to help make their experiment a reality by seeking guidance from CancerCare Manitoba.

"It was a great idea they came up with and we wanted to make it possible," added Nickel. "The Argyle kids took off with it and worked with their classroom teacher Lesley Nesbitt Fuerst."

"Competition is good and it's healthy, and keeps you on your toes," said Nickel.

THE LOCAL PARTNERS THAT MADE THE EXPERIMENT TO GO TO THE ISS POSSIBLE:

Interlake School Division

Government of Manitoba

Manitoba Aerospace Association

Manitoba Aerospace Human Resources Council

Aerotech Herman Nelson International

Bristol Aerospace (a division of Magellan Aerospace)

Boeing

StandardAero

Acsion Industries

Acetek Composites

Manitoba Hydro

Allied Wings, Canada Wings

Mr. Alfonz Koncan

Canadian Space Agency

Oak Hammock Marsh

Fort Whyte Center

Central and Arctic Region, Fisheries and Oceans Canada

Bedford Institute of Oceanography

Kinesiology and Applied Health, The University of Winnipeg

Pembina Trails School Division

Seven Oaks School Division

Mr. Orville Procter

Good Turf Garden Centre

Ricard Farms, Ltd.

Cancer Care Manitoba

Shelmerdine

Manitoba Agriculture, Food and Rural Initiatives

University of Manitoba Human Nutritional Sciences

University of Manitoba Food Science

Straight to the Point Community Acupuncture

Royal Bee Farm of Canada

Cornelia Bean

The SSEP is the first pre-college STEM education program that is both a U.S. national initiative and implemented as an on-orbit commercial space venture.

The Smithsonian National Air and Space Museum, Center for the Advancement of Science in Space (CASIS), Carnegie Institution of Washington, NASA Nebraska Space Grant Consortium, and Subaru of America, Inc., are National Partners on the Student Spaceflight Experiments Program.

Chapter 7 – Teaching Excellence Award

"Here is why I am in Ottawa! So honoured to be a part of this 20th anniversary of the Prime Ministers National award winners . Hardest secret to keep since Sept 4th" (FB Nov 20, 2013).

Prime Minister
Stephen Harper
congratulating Maria
on her award. (Photo
credit: Jason
Ransom, 2014)

(Photo credit: Jason
Ransom, 2014)

On behalf of the Government of Canada
and the constituents of Selkirk-Interlake,
James Bezan, Member of Parliament
is pleased to acknowledge and congratulate

Maria Nickel

On her presentation of the 2013 Prime
Minister's Award for Teaching Excellence
and Excellence in Early Childhood Education

Presented
November 21, 2013

James Bezan
Member of Parliament
Selkirk-Interlake

"PMO Awards day. I'm about to go into the Govenor General's home at Rideau Hall cool." (FB Nov 20, 2013).

"My autographed picture given to me from Canadian Astronaut Dr. Robert Thirsk on my PMO Award, he is so nice — at Delta Ottawa City Centre" (FB Nov 20, 2013

ROBERT THIRSK   Canadä

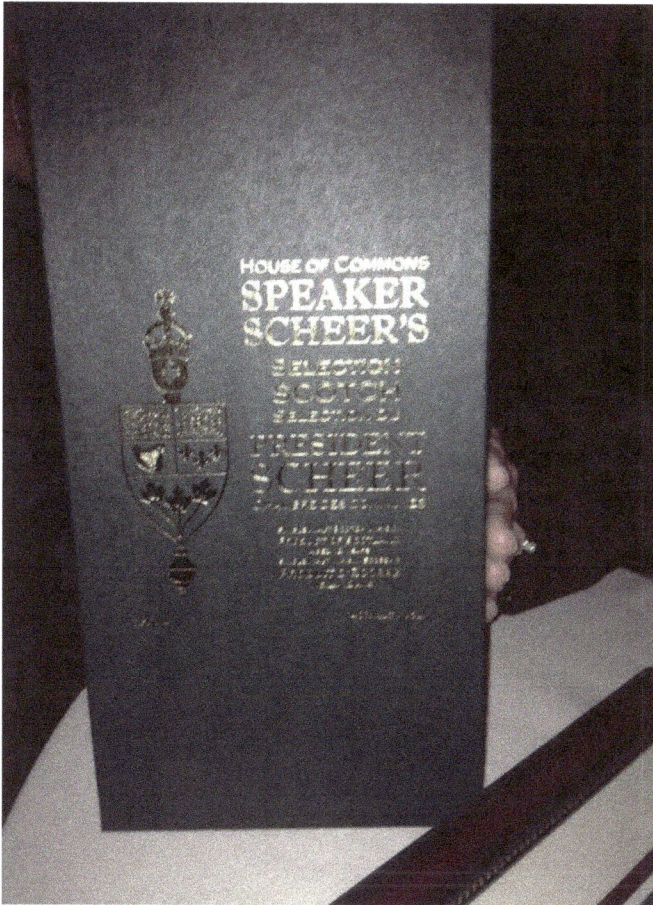

Look what I won at my Prime Minsters Award of Excellence event in wpg [Winnipeg] at the Fort Gary Hotel in the Crystal Ballroom; Speaker of House of Commons 12yr old scotch. Apparently, you can get it if you are invited to the speaker of the house's residence for a tasting, or a visiting dignitary, head of state or in the parliamentary restaurant if you are an MP for $75." (FB Apr 10, 2014).

"Just got this at a school assembly today from PMO in Ottawa! WAY WAY COOL. It was flown in space on the last shuttle mission on Atlantis" (FB June 19, 2014).

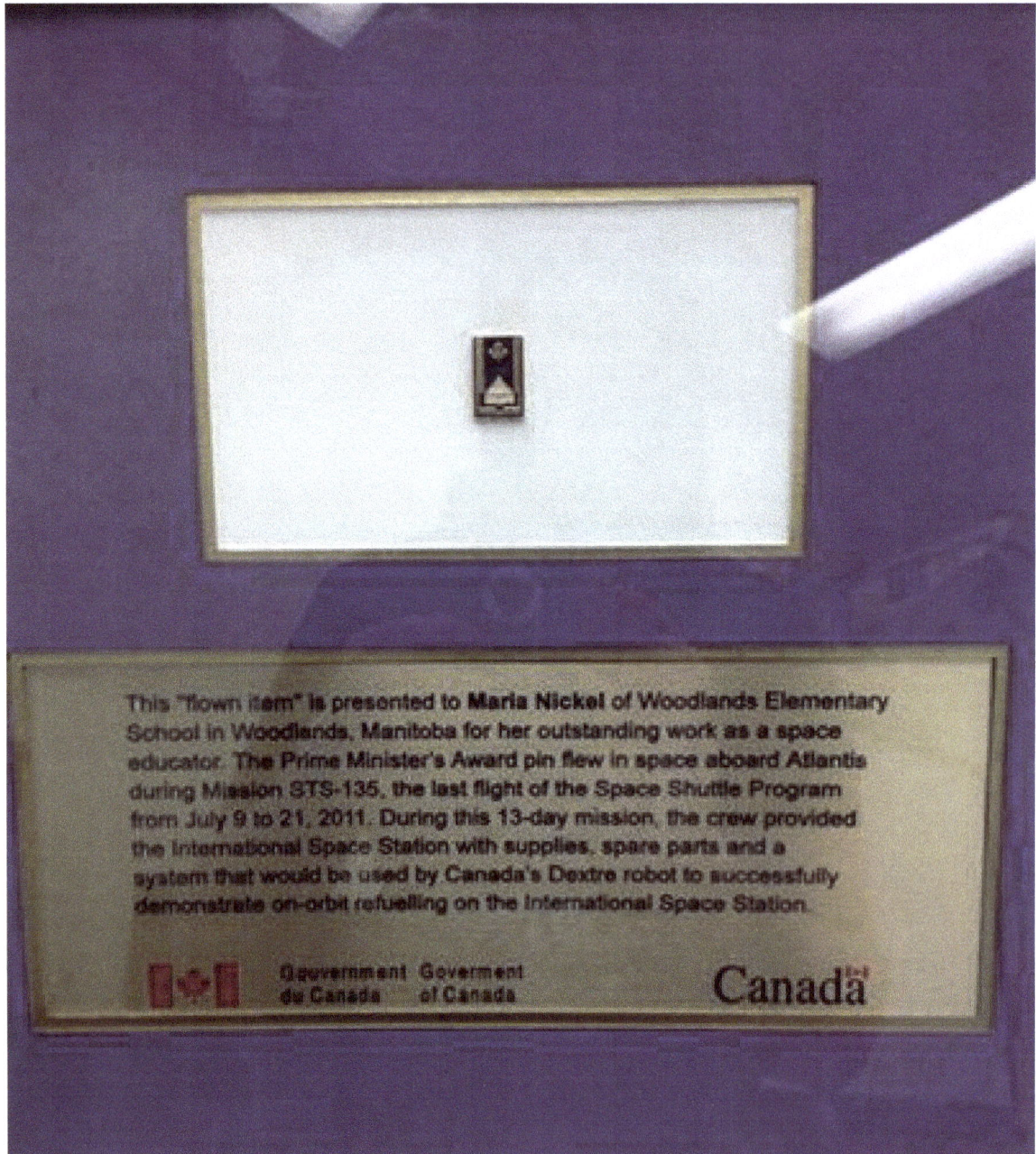

This "flown item" is presented to **Maria Nickel** of Woodlands Elementary School in Woodlands, Manitoba for her outstanding work as a space educator. The Prime Minister's Award pin flew in space aboard Atlantis during Mission STS-135, the last flight of the Space Shuttle Program from July 9 to 21, 2011. During this 13-day mission, the crew provided the International Space Station with supplies, spare parts and a system that would be used by Canada's Dextre robot to successfully demonstrate on-orbit refuelling on the International Space Station.

Gouvernement Goverment
du Canada of Canada

Canada

Maria with husband Bob in Ottawa

OMG what an emotional day for me and my hubby Bob. So much has filled me up and I have so much to say, but just overwhelmed right now. From all my fellow winners coming up and wanting a picture of me in my flight suit I wore proudly in my teacher talk session with media and guests. And now I am referred to as commander by these new found friends and loving what I do with my kids and how in awe they were of our experiment going to the ISS, then getting to meet and have a private chat and picture arranged for me with Canadian Astronaut Dr. Robert Thirsk and him telling me how great my teaching is and the experiment to space and the letter Dr. Jeff Goldstein wrote to the PMO awards on my behalf, JUST WOW!! We talked and giggled and he then asked for a picture with him in my flight suit and with my hubby Bob and he said we need a picture in between the 2 Bob's , how funny is that. Thank you thank you thank you so much for all you super kind , loving words about this award I am to the moon and back on this experience. LOVE LOVE LOVE YOU ALL!! XOXOX (FB Nov 20, 2013).

A well-deserved award Maria. Your school, division, city and province should be proud of you. You have accomplished much for one so young.

Congratulations.

Chapter 8 – The Royals

Woodlands School science and technology teacher Maria Nickel with Brant-Argyle students Ryan Petricig, Avery Good and Ethan Enns, get a close look at their experiment upon its return from the ISS.

(Photo Credit: Stonewall Argus & Teulon Times, May 9, 2014)

On May 21, 2014, they had a meeting with the Prince of Wales (Prince Charles) and Carmilla the Duchess of Cornwall.

Brant-Argyle School became the first Canadian school to have an experiment travel to the International Space Station. goo.gl/lyroKS

The Brant-Argyle School students, Ryan Petricig, Avery Good and Ethan Enns, first place winners along with second and third place winners from Woodlands Space Knights Space Club awaited the arrival of Charles, Prince of Wales and Camilla, the Duchess of Cornwall. These students were accompanied by Prime Minister and Mrs. Harper, the students' teacher Leslie Nesbitt Fuerst, Brant-Argyle principal Laura Perrella, CancerCare Manitoba researchers Dr. Ellen Mckenzie-Matwiy, Dr. Elizabeth Hensen and Dr. Dan Getz, space systems engineer Phil Ferguson and Woodlands School Science and Technology teacher and Director of the Student Spaceflight Experiment Program Interlake, Maria Nickel.

. Maria with her dad and mom. "We're ready for the Royals! Go for launch — at my house just before we left for Stevenson Campus hanger for a Royal visit." (FB May 21, 2014).

Interview with Karen Roznik CTV Morning live — at Red River College - Stevenson Campus Aviation & Aerospace. (Photo by Maria's parents).

Maria waiting for Royals with Phil Ferguson from Magellan (space systems engineer who assisted with SSEP) and the first, second and third place finalists at Red River College, Stevenson Campus Aviation &Aerospace

The Duchess of Cornwall received flowers and asked questions of the students about their experiment to the ISS. (Photo Credit: ©Winnipeg Free Press, Wayne Glowacki)

"Meeting Canadian Space Agency President General Walter Natynczyk (right) & with my parents Stan & Luba Sirdar — at Red River College - Stevenson Campus Aviation & Aerospace. (FB May 21, 2014)

Following the Royals' visit Maria said that the Canadian Space Agency president, General Walter Natynczyk was impressed with the students' science and space knowledge presentation to the Prince of Wales and the Duchess of Cornwall.

"After all of the amazing things that happened today, I also wished I could have had my Baba & Dido and Grandma there to see the Royals with me being 3rd gen. Canadian , first female to graduate high school & university on both sides (they came from poor Ukrainian farming families in the Ukraine) this would have been truly a special moment for them to witness that their first grandchild got to do. This was all for you!! miss you and I know you were with me in spirit — feeling blessed." (FB May 21, 2014).

"CTV interview post royal visit, awesome job by reporter (even got my mom and dad in the shot) (FB May 21, 2014). For the interview see goo.gl/1ODWrk

Maria's space girls, second place winners (on the left) and third place winners (on the right) being interviewed by Beth MacDonell (middle) from CTV at Red River College.

Third place winners of the Student Spaceflight Experiment Program experiment from Mr. Enns class on display for the Royals.

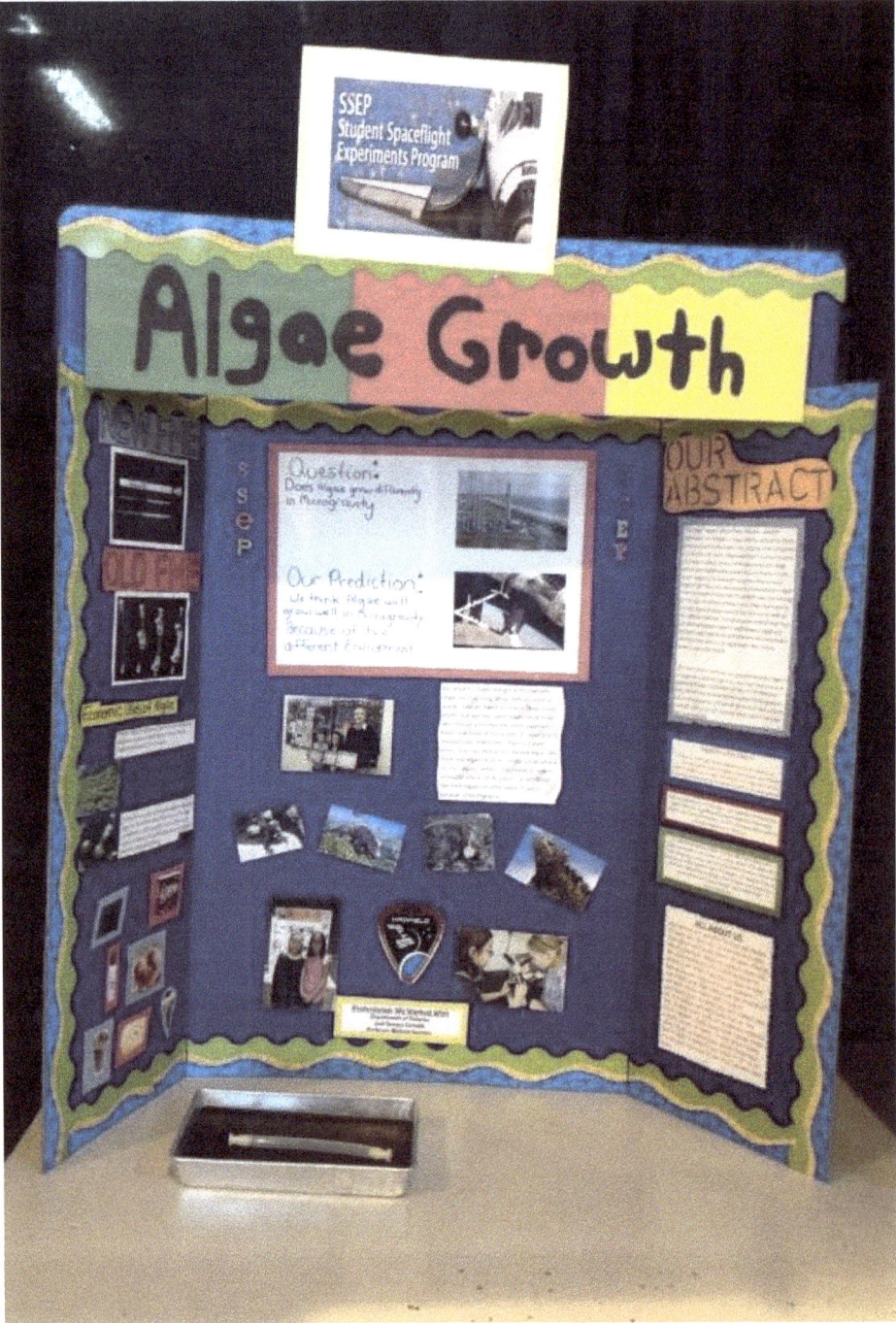

Second place winners of the Student Spaceflight Experiment Program from Maria Nickel's class on display for the Royals.

The winning experiment display of the Student Spaceflight Experiment Program from Lesley Nesbitt Fuerst's class shown to the Royals.

Maria's Interview following the Royal Visit see goo.gl/8l5MBr

"Well there's something you don't get every day. I got an email from the Canadian Space Agency president General Natyncyk about our visit and the great pics I took of him with my family and he is going to be sending me a correspondence for all the experiment kids and a special CSA coin for them & their hard work. I am so excited to get this! very cool." (FB May 26, 2014).

Harold Enns, Maria Nickel, Leslie Nesbitt Fuerst, the third, second and first place students all received this special CSA coin.

What an honour it must have been for the teachers, children and sponsors to have had an opportunity to meet and to speak with Prince Charles and Carmilla, Duchess of Cornwall. What an honour indeed!

Chapter 9 – Maria the Baker

Maria was chosen as one of five finalists to take part in the Kraft Philadelphia Cream Cheese Canada Cheesecake of the year contest. She got to fly to Toronto on July 29, 2014 to compete. The participants were narrowed down from 200+ recipes, to 30, 20, 10 and then to 5 finalists. She was honoured to have had the opportunity to compete.

Some of Maria's other creations:

50th wedding anniversary of my husband Bob's uncle and auntie

VIA Rail Retirement Cake for my sister in law's father in laws' retirement from Via rail as a conductor for 40 years, Via rail featured it in a newsletter.

Winnipeg Blue
Bomber Alumni
Golf Classic cake
(Half white
chocolate cake and
half Guinness
chocolate cake)

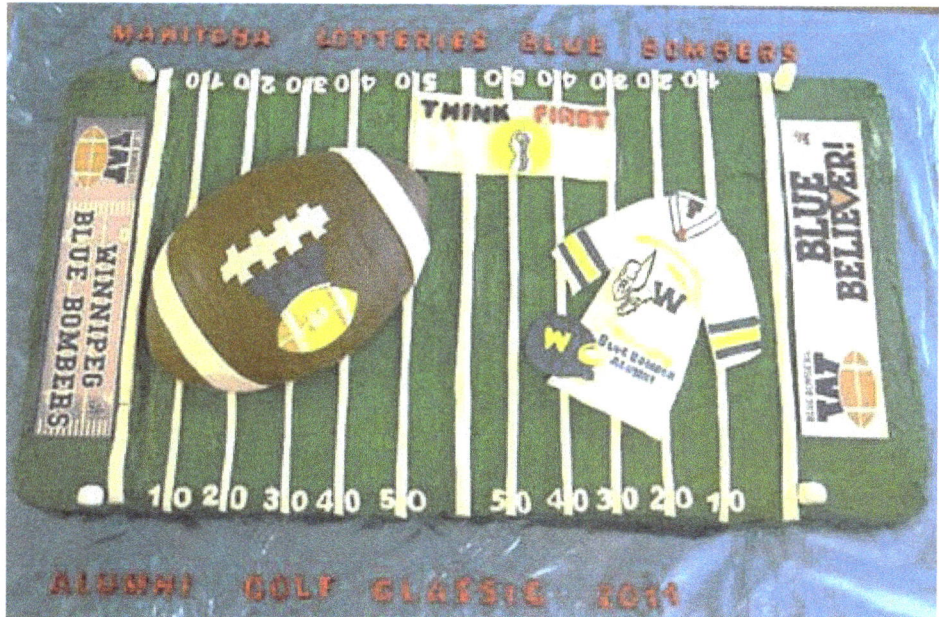

The Darth Vader cake was for Maria's dad's
birthday (he took Maria to see *Star Wars*
when she was a kid and dad loves the
movies). Maria also snuck in her brother Nick
in the cake since he and his dad were close.

Birthday cake made for Maria's cute nephew Blake's
first birthday party in July 2014.

Mom's and Dad's 40th Wedding
Anniversary, June 9, 2009

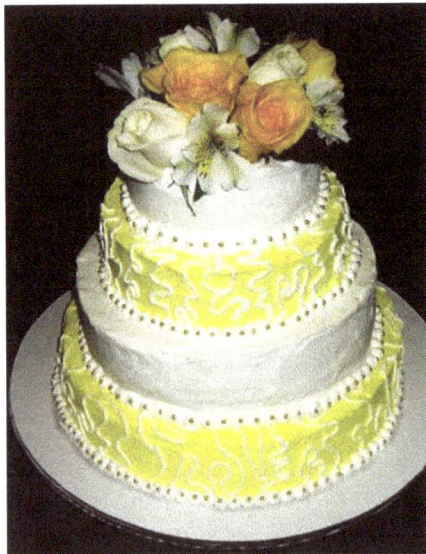

Mary's parents 50th Wedding Anniversary. Mary is Maria's high school best friend They went to school together and are still friends. She asked Maria to make the cake for her parents 50th wedding anniversary. Mary is also a Grade 2 teacher.

Grade 8 Graduation Cake June 26, 2012

Whole cake at home before nerve-racking transport to school for STS - SPACE RIOT 1 March 27, 2011

"I am so VERY VERY VERY EXCITED! I am one of the 5 finalists in this cheesecake competition by Kraft Philadelphia Cream Cheese Canada Cheesecake of the year contest. I get to fly to Toronto July 29 - 31 to bake and compete for the title & $5000. I am so excited and going to practice lots." (FB July 11, 2014).

This picture of Maria's cheesecake was taken in her home. She made the same cheesecake for the competition.

Her cake is made with a chocolate hazelnut crust, a hazelnut cream layer and topped with a Chocolate ganache. It is garnished with raspberries and whole hazelnuts.

Although Maria did not win the competition, she made new friends and enjoyed a wonderful experience.

"So far this had been a unique once in a lifetime chance! I am happy I got to try & play baker at Kraft!" (FB July 30, 2014).

Chapter 10 – Who is Maria Sirdar-Nickel?

June 22, 1996, Maria Sirdar married Robert Nickel. They danced to "Long as I Live" by John Michael Montgomery.

"I was an extra in the Jack Layton film. I played a politician at a lunch bar, a factory worker, a homeless person, and walking in public in China town. I got to wave and acknowledge Jack as he jogged by the hotel. So excited to see if I may be in it. I'm such a geek. March 10th on CBC TV http://www.chrisd.ca/2013/01/27/jack-layton-biopic-cbc-film-premiere/" (FB Jan 27, 2013).

"I am SUPER SUPER SUPER CRAZY EXCITED. I have been asked by NASA education to help edit and pilot some new space, science and math curriculum. BOOOOYA baby!! I am going to get me to space yet!!" (FB Feb 21, 2013). "Well my nephew loved his present. The look on his face was awesome, he was stunned at what I got him. he loves to read , so I thought of him at the time I got this at my first space academy with team Exploration. We had the honor of listening to Homer Hickam, who spoke about his journey to working for NASA and helping with the rocket program and building the shuttle boosters to space and his adventures as an author. I have been saving autographed copy of "Rocket Boys" (movie is October Sky about his life story about rockets and his journey to NASA) from Homer to my nephew since then (2009) for his 13th birthday when he could read it. I told him the story behind it and he was in awe and held it and read the back and held it and said   he is going to read it. It's not a wack of money , but a little bit of something unique." (FB Feb 24, 2013)

"OMG OMG. Homer Hickam (right) [see the movie "October Sky"] just replied to my message about the book I gave my nephew. I am so thrilled to have chatted with the original Rocket Boy!! Wowza. What a nice guy!!" (FB Feb 24, 2013).

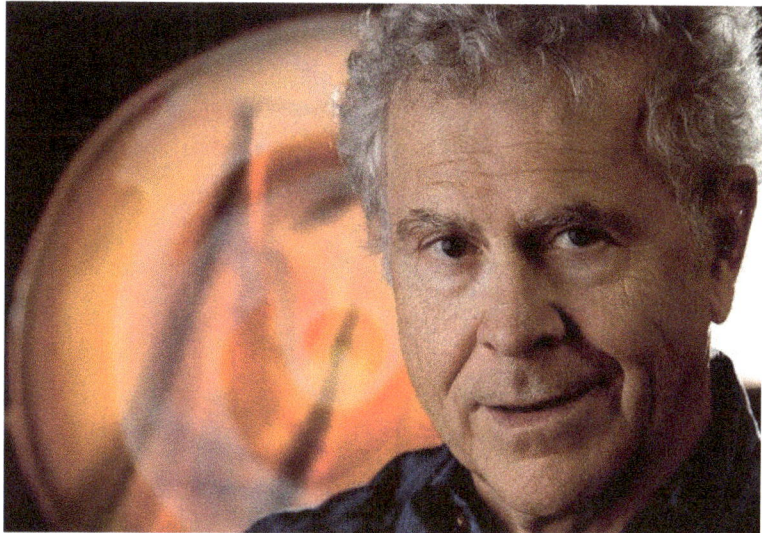

Marian Gilmore writes, "I met Maria at space camp, but didn't get to know how dedicated and passionate a teacher she was until later. I have attended several conferences with Maria to see her love of teaching shared to teachers around the world. I have also read about the many awards and acclamations that Maria has won over the last few years. Maria is the teacher you wish your children would have and the teacher you wish you had." (Marian "Mare" Gilmore, Vice President, Inspiring Educational SySTEMs, Inc. and Former Director of Education at the U.S. Space & Rocket Center Aug. 24, 2014).

"Space Riot A SUCCESS. CTV NEWS AT 6 pm TO Showcase space science activities and honoring Cmdr. Hadfield. Marilyn Maki interview at noon done (was cool) and interview with Argus as well." (FB Mar 1, 2013).

"Many were asking what we did during space riot. Here we go: AIRPLANE DESIGN, HELPICOPTER DESIGN AND TARGET LAUNCH, CREATURE CREATIONS USING HOMEMADE PLAYDOUGH AND GOOGLY EYES, ORBITING AROUND THE EARTH CRAFT (on a paper plate was earth and orbiting around it was you as an astronaut, a rocket and an alien in a ship), FINGERPUPPET ALIENS made out of pompoms , googly eyes and pipe cleaners, SPACE WALK RELAYS on an obstacle course to train like an astronaut, PLANETARY TOSS (bean bags into hoops with planets at each one with a point value for little guys and Basketball hoop for big kids), METEOR SHOWER HIT (blindfolded kids and had to land your asteroid on a planet or a star on a giant board with all those object on them), WATER MAZE (had to guide a water droplet through a maze with your pipette) & PENNY DROPS (guess and try how many drops of water fit on a penny in heads and tails, a bit of surface tension experiments) thanks Kim Astromovich for that one. WHAT AN AWESOME AWESOME DAY OF FUN. Hmmmmm maybe I should present my riot at SEEC 2014. Oh the fun for teachers" (FB March 2, 2013)

"So very proud of my students (whose story was replayed on last night's CBC national news) during their whole evenings coverage on the return of Chris Hadfield to earth. They even updated the story on when our experiment flies to the ISS , it leaves on Nov. 11, 2013 barring any delays. SO PROUD!! goo.gl/zEZR9l

"I had a cool event happen Friday night at MTS AGM. Myself and my associate delegates were out for dinner before our next session at 7 pm - 9 pm, when we went to a restaurant near the hotel and the server says to me " Do you remember me Mrs. Nickel, you were at John Henderson right?" I quickly scanned her face and saw something familiar but not sure where, so with a big smile I said, "Why yes I do?" & yes I did go to JH." She then says you did science right and something else? I said yup, and phys ed. " She then says, I was in Gr 7 and then said more names of kids I taught that year, I recognized (that was 13 yrs ago) and then she says " I went into University and guess what? I went and got my biochemistry degree in science and I work for a pharmaceutical company and I am going for my masters in Physicians Assistant, all because of you and your science" I stood, speechless and then congratulated her on your accomplishment and hugged her" She smiled so hard, I thought her face would break /o/o. She then served our meals that night. It was so so so cool to see a student you impacted so positively!! I am humbled right now and have so many goosebumps" (FB May 25, 2013).

"Well so far so good, things are OK! Yay, but that may change, I am covering a class to do 2/3 physed as my 5/6s are on field trip, should be interesting, never done 2/3 gym before, if I had a pool we would be doing swimming lessons. ahhh the lifeguard in me is back again, can't get it out. Hahaha" (FB May 30, 2013).

"You know you are not ready for winter when you go to Ottawa [to receive the Prime Minister's Award for Teaching Excellence] & as you fly in people are still golfing , and I come home go to school wearing winter boots but forget to put them on and go home in runners and now my feet are cold today cause I got no boots till I get back to school hahahahaa" (FB Nov 23, 2013).

"Of course it is - 41 degrees C with the windchill & it is the first day I am trying my knee at my 1st spongee game of the year. Go Courgars !! let's see how we fair, 1 of 2 scenarios: knee is fine or knee is done and am out for rest of yr for spongee!! we shall see" (FB Dec 29,2013).

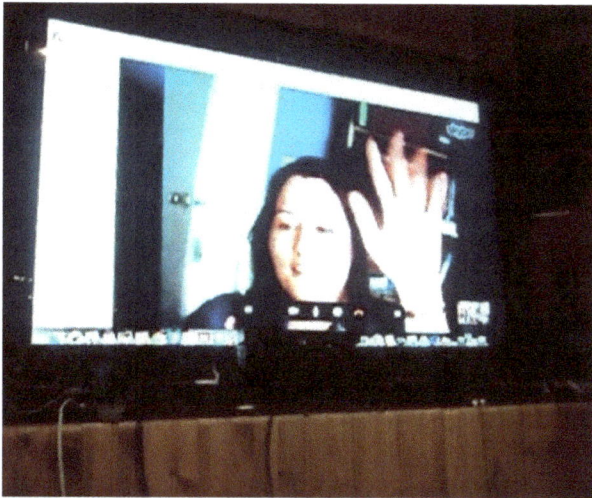

"The AstroNuts Kids Space Club with Brett and his Dad Ray Bielecki in Ontario, Canada. I have been working with them for the past 3 yrs, with various space topics, activities and Skype sessions. I am Skyping with them on this Sunday, super excited, we are going to talk about our experiment going to the ISS and our mission patches already in space right now. so cool!! [E-mail from Maria Nov. 2014 See a brief clip of Maria's Skype with Ray and the AstroNut kids at http://goo.gl/cfVKnJ]

Maria in Winnipeg on Skype with Ray and the AstroNut kids in Newmarket, ON

" Ok Bobby J, you are nuts, this is NOT SAFE!! I don't care how much you want the games in today or yesterday! a -51 windchill is NOT SAFE to play sponge hockey, no matter how warm one dresses & me being goalie is the worst position for cold. Today is not going to be a good day." (FB Jan 5, 2014).

"I made it!! had to go in between games and Sharon Allen & Cory both helped me warm my feet up and I pulled my socks off to use on my hands to rewarm them! it was not too bad for 2nd game, still froze, am now home warming up in my blankie, Bailey's & hot chocolate & pancakes. super effort ladies of spongee: Brenda Bourns I hope your head is ok, that was 2 nasty spills today even with a lid on! Mena, Pauline, Shauna,Sandy, Jan, I know I am forgetting someone, but ya all did super out there!" (FB Jan 5, 2014)

"We are go for launch now Wed Jan 8th at Manitoba Museum in the auditorium near the science gallery at 11:30 am. Just got some amazing pics via our awesome NASA Wallops engineer Catherine, taken Dec 16th when we were there and they were prepping her for launch that is now happening in 2 days. Still cool pics." (FB Jan 6, 2014).

"Off to Orlando FETC major tech conference to do training and awesome PD on tech for teaching. Can't wait to learn so much! Going to miss my students , space club kids, hubby, buddies , Spongee gals, & family! Bk in a week hopefully with some free tech goodies I hope to win! (FB Jan 26, 2014).

"I'm loving this tech conference FETC here in Orlando but I think I brought Nov weather here LOL from home! I got nothing to wear here SO COLD I had to buy another jacket I so brought the wrong clothes, who woulda thought u don't bring enough pants to wear for here. I got capris , shorts & dresses not worn. Only 1 pr of shorts worn & that was Monday in Cocoa beach & the rest of the time is cold winds & rain like our weather in late Oct /Nov. I am stunned at this out here! Last day here, hope my sessions r just as amazing as yesterday. — feeling wonderful." (FB Jan 31, 2013).

"The NASA curriculum I was working on with my test group (Space Knights Space Club) is now posted on their [NASA] site. Go to the introductory page to see my name along with Jacqui Flowers , Becky Loy, and Amy S. Bartlett. Really cool experience LINK TO NASA AREA THIS IS FOUND ON: http://www.nasa.gov/audience/foreducators/best/activities-technology.html (pg 8 of document0 LINK TO WORKBOOK!" goo.gl/JLQuBB (FB Feb 15, 2014).

"I just got the most wonderful comment from a super amazing parent of a student I taught who had to go to another school last week for sports reasons. I loved this student of mine, she is so talented and amazing & I hope to see her in the Olympics one day, she is that good! Anyhoo this is what I got from this parent: 'My daughter told me what you said about following her dreams and setting goals. Thank you Maria Sirdar-Nickel for being such a wonderful teacher!! It is that heart and compassion that makes you a wonderful teacher for all your students!' She totally made my week with that comment, I got a bit teary eyed afterschool today sniff sniff!! Time to go home early today for once." (FB Mar21/14)

"Off to MTS for Collective Bargaining mtgs with all provincial bargainers 9 - 4 pm so that we can help our teachers collective agreements be positive in benefits which benefits our kids in the end . Happy teacher = happy kids & parents! We want teachers to stay in the profession not run from it because the working conditions are so intolerable they take other employment and say this profession is not worth it. As a teacher member I am trying to help my colleagues and do my part for the benefit of all! Happy Sat.! One more week till spring break woot woot" (FB Mar 22/14)

"LAST DAY OF SCHOOL! So relieved for spring break, to catch up on entering marks, prepping for AAIMS day with space club & winning experiment ISS kids 1st , 2nd & 3rd place details, to send out some thank you cards to NASA HQ when we visited Washington, the Cdn Consulate staff & CSA consulate rep Bill Mackey, MARS engineer Christine for her help , perogy making, babysitting my godson & neice Sat.,some movie time I think and shopping NOT at 8 pm at nite for a change. So looking forward to some sleep in time. Few hours yet. Best part today is my excited Gr 5/6 class to watch Apollo 13 to help prep for their Moon Base building design challenge for their final unit project. Going to be super great to see how much they are going to love it! (FB Mar 28/14).

Chris Hadfield. Signs his book, "An Astronaut's Guide to Life on Earth" in Winnipeg in 2014.

Meeting Canadian Space Agency President General Walter Natynczyk at Red River College - Stevenson Campus Aviation & Aerospace.

Manitoba Provincial Baseball Slo-Pitch Champs (Maria is on the far left)

Maria won a driver for the Women's Longest Drive in the Ellen Chlan Memorial Golf Tournament. All money goes to the Rehab Centre for Children Hospital in Winnipeg.

Did you know that Maria was a trainer for the Atlanta Braves one summer?

"Atlanta Braves Spring Training 1997. My days as a guest Atlanta Braves trainer — in West Palm Beach, FL, United States. (FB Mar 15, 1997).

Family has always been an integral part of Maria's life.

Front Row: Luba and Stan (parents)
Second row L to R: Julianna (youngest, 18 yrs difference between her and I , she was 7 yrs old at the time as my flower girl at my wedding in 1996), Nicholas (4th oldest), John (3rd oldest)
Third Row L to R: Anna-Marie (5th oldest), me (oldest), Chris (2nd oldest)

All Maria's cousins and brothers and sisters at the house of Baba and Didos

Maria and her family skating at Kildonan Park, Winnipeg

Maria's 18th Birthday with her family

Maria (age 21) and her husband Bob are teaching Maria's little sister Julianna how to skate.

Chapter 11 – Maria's Advice to Students Regarding Career Options

To be an educator requires years of training and like any profession, the training continues after the university courses are completed and the degrees earned. An elementary school teacher is trained in all subjects and is expected to teach all subjects. A secondary school teacher studies one subject and teaches that one subject.

To be an educator, make sure that you like young people. You will need unlimited patience and ensure that you are a good listener. You have to believe that young people are our future.

In elementary school, get involved in extra-curricular activities like sports, class newsletter, or the year book committee. Study hard and complete all assigned work to the best of your ability. Do the same in high school. If you are shy, join the drama club. It will gradually rid you of your shyness and you will have fun in the process. Also in high school, choose a subject that interests you the most and continue to take more courses in this same subject (English, Science, Geography, History, etc) in University. Your first degree upon graduation will be a Bachelor's Degree in Arts, Science, or Physical Education to name a few.

Once you have earned your first degree, you will attend Teachers' College to earn a Bachelor of Education. It is at Teachers' College that you will decide if you want to teach elementary or secondary school. This training supplies you with the knowledge to teach your subject(s), prepare lesson plans, prepare fair tests, and how to deal with a plethora of individuals at different academic and social levels.

You might want to pursue a Master's Degree while teaching. Once this degree is earned, positions like Vice-Principal, Principal and Guidance Counsellor, for example, become available.

Education is a most rewarding career, but it has its drawbacks like long hours, seemingly endless marking of students' assignments and tests, and never enough time to cover the curriculum. On the positive side are appreciative students, summer holidays and camaraderie among one's peers. The pay is respectable and the benefits are competitive."

Teaching is the gift of touching thousands of lives for the benefit of humanity. Teaching goes beyond just the textbook. It is relationships that last a lifetime that bring me the most joy. It really hits me the most when I have former students now in university or in careers that come up to me and say "I remember that fun lesson when we went on that Galapagos adventure in science and ELA" or "You were my favourite teacher. You always had the coolest experiments and activities to do. I hope to be a great teacher just like you one day". Those are the impacts you can have on kids in teaching. I do it because I care and want the best for all kids of all abilities. A good teacher can shine brightly to inspire their students to levels of greatness they never thought possible and I strive to do that each day I walk into my classroom.

Chapter 12 – Summary of Achievements

Maria has had a number of exciting things happen in her life. She

- won the National Prime Minister's Award for Teaching Excellence;
- was nominated by her school board for the Xerox Innovators Award;
- was the keynote speaker for many conferences including the Manitoba Aerospace Assoc;
- helped to send an experiment to the International Space Station;
- got a standing ovation at the Manitoba Teachers' Society AGM from all 400 delegates;
- met Prince Charles and Camilla, Duchess of Cornwall to talk about her space club and SSEP;
- went to conferences with beloved friends to learn about space to implement activities in the classroom;
- met and had pictures with not only one but three Canadian Space Agency Astronauts, namely: Chris Hadfield, Dr. Robert Thirsk & Jeremy Hansen;
- met with CSA President Gen. Walter Natyncyk;
- met NASA Astronaut Jerry Ross USA an inductee into the Astronauts Hall of Fame;
- was nominated for Manitoba Heroes Award;
- got to take her parents to meet the Royals;
- was nominated for the Cherri Brinley Award (JSC NASA Houston, TX)
- featured speaker STAO, November, 2014

Maria's Honours and Awards to September, 2014

Winner Prime Ministers National Award Teaching Excellence 2013

Prime Ministers Office, November 2013 (http://www.pma-ppm.gc.ca/eic/site/pmate-ppmee.nsf/eng/wz02086.html).

Nominated Manitoba Heroes Award 2014 Gala. Our Manitoba Heroes 2014. This award is a celebration of Manitoba's everyday people, doing extraordinary things.

2014 Finalist Philadelphia Cheesecake of the Year Contest, Kraft Real Women of Philadelphia. July 2014. was in the top five to compete in a cheesecake bake off at Kraft Foods Ltd at the test kitchens in Toronto, July 28 - 31, 2014. Maria Nickel's entry was Chocolate Hazelnut Cream Cheesecake. http://realwomen.phillycanada.com/blog/post/Cheesecakes-Around-the-Country-Where-Our-Finalists-Are-From and ttp://realwomen.phillycanada.com/blog/post/Our-Top-5-Finalists-Revealed-Cheesecake-of-the-Year.

Nominated for the Annual Cherri Brinley Space Science Educator Award 2014, Johnson Space Center Houston at NASA, February 2014. Cherri, a fourteen year SEEC veteran, shared her passion for Space Science and inspired so many educators and students around the nation.

Nominated for the Amgen Award for Science Teaching Excellence 2014. Amgen Inc., January 2014 This award recognizes extraordinary contributions by educators across the United States, Puerto Rico and Canada who are elevating the level of science literacy through creativity in the classroom and motivation of students. Amgen Inc., is a biotechnology pioneer, that markets important human therapeutics for serious illnesses.

Nominated& Finalist Manitoba School Boards Association Xerox Innovators Award – 2013. Manitoba School Boards Association & Xerox March 2013

Nominated& Finalist Board of Trustee Interlake School Division for the Manitoba School Boards Xerox Innovators Award – for the Student Spaceflight Experiment Program

The Right Stuff Teacher Medal winner 2009, Honeywell Hometown Solutions Space Academy for Educators June 2009

Scholarship Winner $1500 US Advanced Space Academy for Educators –2011. Honeywell Hometown Solutions June 2011

Scholarship Winner $1500 US: Space Academy for Educators –2009. Honeywell Hometown Solutions

Manitoba Government – Education June 2013, Grant winner - $1000.00 – 2014 – 2015; Grant winner - $1000.00 – 2013 – 2014; Grant winner - $1000.00 – 2012 – 2013; Grant winner - $900 – 2011 – 2012; Grant winner - $1000 - 2009 – 2011. Manitoba Education has established Scientists in the Classroom grants to support teachers in their efforts to collaborate with scientists who agree to engage with students in the classroom. Grants are awarded based upon demand and availability.

 Grant winner - $1000. Science Teachers Assoc. Of Manitoba, Science Teachers Assoc. Of Manitoba, September 2011. Grant money secured for Woodlands Elementary Space Knights space club space science activities. Activities involved : science, math, engineering , and technology integration for students Grades 6 - 8.

Grant winner - $900. Canadian Space Agency Educators Conference. Canadian Space Agency August 2010.

Nominated Prime Ministers Award for Teaching Excellence Science Education, Prime Ministers Office, October 2012, October 2011.

Nominated by the Woodlands Elementary Parent Advisory Council for the : Minister of Education Manitoba Excellence in Teaching Award for Group Collaboration for the program Global Voyageurs Trek program for students gr. 4 – 7. Special Area Groups Presenter for the Manitoba Science Teachers Association, SAG, October 2012

Winner, Minister of Education Manitoba Excellence in Teaching Award for Group Collaboration, Minister of Education Government of Manitoba, April 2011

Featured Speaker, Science Teachers Association of Ontario (STAO) presenting "Toys in Space." Video on YouTube of Maria's Talk at STAO 2014 "Toys in Space" goo.gl/ofue8N

Interests

Baking and cooking and entering any contest (winner of several prizes). Entering cooking and baking contests (winner of several prizes), baking for my students and Gr. 8 grad cake, entertaining, horseback riding, reading science fiction, skyping with friends, spending time with family and friends, playing softball, sponge hockey, volleyball, basketball, beach volleyball, swimming, travelling, planning activities for my Space Club Kids Gr. 6 – 8. .

Maria is assisting/consulting on the second Canadian experiment to space from Grade 6-7, McGowan Park Elementary, Kamloops/Thompson #73 School District., BC. The Teacher Facilitator is Sharmane Baerg" (Resource: http://ssep.ncesse.org/communities/selected-experiments-on-ssep-mission-6-to-iss/) Unfortunately the Antares rocket blew up on Oct. 28, 2014 just seconds after launch. The experiment will be re-launched in January 2015.

Parent Testimonial: "Bradley came home with many wonderful life stories from Mrs. Nickel....great teachers like you inspire our children to go out into the world with knowledge, confidence and respect! Thank you for all the extra time you gave over the years to both my boys!! " Lisa Lillies, Woodlands Elementary School parent.

Maria is a creative, resourceful, dedicated and a passionate educator. She has, in her short life, accomplished so many wonderful things for her students and her colleagues. I have a feeling her journey is just beginning and I wish her continued success. Stanley R Taylor.

The colour of the front and back cover of the book is one of the colours from the University of Manitoba, Brandon University, and the Coat of Arms for Manitoba.

About the Author

Stanley R. Taylor taught for the Toronto Catholic District School Board for 23 years. He retired in 2001 and started doing workshops for students, educators and within the community for Scientists in School (www.scientistsinschool.ca).

Stan has given workshops on how to build his pneumatically controlled Canadarm for the Science Teachers' Association of Ontario (S.T.A.O.) annual conference, the Ontario Association of Physics Teachers annual conference and the Science Exploration Educators Conference (NASA). Stan's Canadarm is the signature toy in his book, *Taylor's Pneumatic Toys*.

Stan has published articles in *Crucible* and *Elements* (the two e-journals of the Science Teachers' Association of Ontario), *MetroVoice* (newsletter of the Ontario English Catholic Teachers Association), *Canadian Teacher*, *Uxbridge COSMOS* and the *Fredericton Gleaner*.

Stan is a member of the following organizations:

- Retired Teachers of Ontario
- Professional Writers Association of Canada
- Writers Community of Durham Region
- Montreal Press Club
- Program Committee of S.T.A.O.
- Royal Astronomical Society of Canada
- Scientists in School